About the Author

Dr. Devesh Thakur has been serving as Faculty Member at the College of Veterinary and Animal Sciences, CSKHPKV Palampur, for the last 15 years. He holds MVSC and PhD degrees from the esteemed Indian Veterinary Research Institute. Dr. Thakur's academic journey has been marked by excellence, exemplified by securing top rank in the Indian Council of Agricultural Research Postgraduate examination.

Specializing in Veterinary Extension Education, Dr. Thakur's postgraduate work focused on the crucial intersection of gender factors in dairy farming. His commitment to education, research, and extension activities spans over 15 years, during which he has made substantial contributions to the field. Dr. Thakur's prolific publication record includes over 72 research papers, 178 popular articles, 13 technical bulletins, and four books. His previous works encompass critical topics such as Gender Factor in Access to Livestock-based Information in India, Livestock Entrepreneurship, Dairy Entrepreneurship, and the Use of Social Media in Agriculture and Animal Husbandry. Beyond traditional publishing, he has ventured into the digital realm, creating a comprehensive 10-hour Online Course featuring 86 videos on Public Speaking, Communications, and Interpersonal Skills.

As a Principal Investigator, Dr. Thakur has led eleven action research projects funded by prestigious national and state agencies, showcasing his expertise in areas such as NIF-DST India, NABARD, HIMCOSTE, National Commission of Women, and ICAR-IASRI, New Delhi. Dr. Thakur's commitment extends to the grassroots level, where he has conducted numerous training programs, Model Training Programmes, and workshops. Funded by key entities like the Agriculture Skill Council of India, National Commission of Women, Ministry of Agriculture (Government of India), NIFTEM, and ICAR, these initiatives have targeted veterinarians, para-veterinary staff, extension officers, and farmers, providing invaluable insights into Dairy, Sheep, Goat, and Poultry Farming.

The culmination of his expertise is evident in his latest venture, "Gender Perspectives in Animal Husbandry" Crafted with a keen understanding of diverse learning needs, the book offers a range of question types, making it an invaluable resource for students, educators, researchers, and professionals navigating the complex relationship between gender dynamics and animal husbandry.

Gender Perspectives in Animal Husbandry

Devesh Thakur B.V.Sc.& A.H., MVSc (IVRI), PGDRD, PhD (IVRI)
Assistant Professor
Department of Veterinary and AH Extension Education
DGCN College of Veterinary and Animal Sciences
Himachal Pradesh Agricultural University
Palampur 176 061, Himachal Pradesh, India

NIPA® GENX ELECTRONIC RESOURCES & SOLUTIONS P. LTD.
New Delhi-110 034

NIPA® GENX ELECTRONIC RESOURCES & SOLUTIONS P. LTD.

101,103, Vikas Surya Plaza, CU Block
L.S.C.Market, Pitam Pura, New Delhi-110 034
Ph : +91 11 27341616, 27341717, 27341718
E-mail: newindiapublishingagency@gmail.com
www: www.nipabooks.com

For customer assistance, please contact
Phone: + 91-11-27 34 17 17
Fax: + 91-11-27 34 16 16

Print ISBN: 978-93-58876-39-0

ebook ISBN: 978-93-58878-64-6

NIPA® also publishes books in a variety of electronic formats. Some content that appears in print may not be available in electronic books, and vice versa.

Composed and Designed by NIPA®.

Preface

Welcome to "Gender Perspectives in Animal Husbandry," an in-depth exploration of the complex relationship between gender dynamics and livestock production. This book is born out of a profound passion for understanding and addressing the nuanced intersections of gender and veterinary science. Each chapter provides valuable insights that enhance our understanding of how gender influences and shapes the landscape of livestock production.

Chapter 1, "Exploring the Intersection of Gender and Livestock Production," lays the foundation by examining the fundamental concepts of gender and sex and their implications for animal husbandry. Through real-life examples, we unravel the complexities of these concepts and their profound impact on societal roles and expectations.

In Chapter 2, "Women's Role in Livestock Production," we highlight the vital contributions of women in rural communities and the nation's economy. We delve into the tasks and responsibilities shouldered by women in livestock care, while also acknowledging the barriers they face.

Chapter 3, "Importance of Gender Sensitization in Animal Husbandry," underscores the critical need for raising awareness and fostering inclusivity within the livestock sector. We explore strategies for promoting gender equality and challenging stereotypes.

Chapter 4, "The Importance of Gender Analysis in Livestock Farming," delves into understanding and addressing gender disparities within the sector. Through gender analysis, we uncover disparities in roles, resource access, and decision-making power.

In Chapter 5, "Gender Mainstreaming in Animal Husbandry," we explore the concept of gender mainstreaming as a strategy for promoting equality. By integrating a gender perspective into policies and programs, we strive to recognize and address the distinct needs of women and men.

Chapter 6, "Gender Budgeting in India," examines the intricacies of gender budgeting and its importance in promoting gender-sensitive policies. We discuss tools and principles to deepen our understanding of how gender budgeting can contribute to more equitable development within the livestock sector.

This book is intended for students, researchers, and anyone interested in understanding gender dynamics in animal husbandry. It serves as a valuable resource for promoting equality within the farming community. I hope this textbook sparks your curiosity and contributes to the ongoing conversation about gender roles in animal husbandry. Happy reading!

Acknowledgement

It is with immense pleasure and profound gratitude that I present this book, "Gender Perspectives in Animal Husbandry," to readers interested in the intersection of gender studies and veterinary science. This project has been a labor of love, fueled by a deep passion for both fields, and it would not have been possible without the invaluable contributions of several individuals whom I wish to acknowledge here.

First and foremost, I extend my heartfelt appreciation to Professor S VN Rao, Former Head of the Department of Veterinary and Animal Husbandry Extension Education at Rajiv Gandhi Institute of Veterinary Education and Research (RIVER) Puducherry. Professor Rao's dedication, expertise, and meticulous editing of six chapters of this book have significantly enriched its content. His invaluable insights and unwavering support have been instrumental in shaping this work, and I am profoundly grateful for his mentorship throughout this endeavor.

Furthermore, I express my sincere gratitude to Dr. Mahesh Chander, Principal Scientist & Former Joint Director, Extension Education at the Indian Veterinary Research Institute, Izatnagar, for his pivotal role in guiding my academic journey. Dr. Chander's mentorship, encouragement, and scholarly guidance during my MVSc and PhD studies have been invaluable in shaping my research interests and academic pursuits.

I am deeply thankful to the Vice Chancellor of CSK Himachal Pradesh Agricultural University for granting me permission to undertake and complete this book project.

Additionally, I would also like to extend my gratitude to Dean & Faculty, Dr. G C Negi College of Veterinary and Animal Sciences, Palampur, for their unwavering support and encouragement throughout the process of writing this book. Their guidance and encouragement have been instrumental in bringing this project to fruition.

The journey of compiling this book has been both enlightening and rewarding, and it would not have been possible without the unwavering support of these esteemed individuals. Their contributions have significantly enriched the content and quality of this work, and for that, I am profoundly grateful.

I extend my sincere thanks to all the readers and scholars who will engage with this book, and I hope that it sparks meaningful dialogue and exploration in the realm of gender dynamics in animal husbandry.

Devesh Thakur

Contents

1

Exploring the Intersection of Gender and Livestock Production Understanding the Difference between Gender and Sex

Chapter Overview

This chapter provides a comprehensive understanding of the intersection of gender and animal husbandry, emphasizing the influence of culture, society, and history on our perceptions of gender and sex. It explores the interconnectedness of these concepts and their significance in shaping roles and expectations of the members of a society in livestock production. This chapter delves into the complex concepts of gender and sex, shedding light on their intricate nature and exploring their implications for animal husbandry. It offers a comprehensive understanding of the biological aspects of sex and the socially constructed aspects of gender. Real-life examples are used to illustrate the significance of several criteria in classifying individuals, as well as the impact of societal expectations, norms, and cultural influences on individuals' roles and behaviours. The chapter also highlights how gender goes beyond the confines of biological sex and is a product of cultural and societal influences, affecting various aspects of life, including tasks in animal husbandry.

SECTION A: THEORY

Sex

In the realm of biology, the term "sex" is multifaceted and extends beyond just sexual behaviour. Instead, it encompasses a range of definitions that delve into the fundamental aspects of maleness and femaleness. In this chapter, we'll explore the primary biological criteria that help us classify individuals as male or female: chromosomes, hormones, and reproductive anatomy.

I. Sex: Categories based on Reproductive anatomy, Chromosomes and Hormones

When we speak of sex in terms of biology, we're referring to the biological or physiological aspects that distinguish males from females. These aspects include the anatomical differences in the reproductive systems of males and females. For instance, males possess testes responsible for sperm production, along with a penis and scrotum. Females, on the other hand, have ovaries that produce eggs, along with a clitoris, vagina, and uterus. Sex, in the context of biology, can also be seen as distinct categories based on genes, chromosomes, and hormones. These factors serve as defining characteristics that classify individuals as either female or male.

Example: Rohit and Rita, an expectant couple eager to know the sex of their unborn child through prenatal testing. During the ultrasound, the doctor informs them that the baby possesses XY chromosomes, indicating a biological male. This determination is made based on the presence of the Y chromosome, which typically leads to the development of male reproductive organs and the production of male hormones.

II. Sex: Traits and Characteristics Resulting from Biological Origins

Another perspective of sex in biology is the traits and characteristics that arise from biological origins. These include traits such as the possession of breasts, having given birth, and the capacity to nurse a baby, which are exclusive to biological females.

Thus, we've explored the multifaceted nature of sex in biology, highlighting how it encompasses reproductive anatomy, chromosomes, hormones, and resulting traits. Understanding these dimensions is crucial for appreciating the complexity of biological sex and its role in our understanding of human biology and diversity.

Gender

In the realm of Sociology and Psychology, the concept of gender has evolved beyond a simple binary classification based on biological differences between males and females. It has expanded to encompass the intricate interplay between cultural, social, and psychological factors that shape the roles and identities of individuals within society. Gender (noun) is derived from the Latin word "genus" referring to kind or race. Gender (noun) is defined as "a kind, sort, or class referring to the common sort of people.

I. Gender: Social Groups or Categories

i. American Psychological Association (APA) says that "gender is cultural and is the term to use when referring to women and men as social groups"

   This perspective underscores the idea that gender goes beyond mere biological distinctions, emphasizing the cultural and societal dimensions that define individuals as part of distinct social categories.

ii. Helgeson (2005) wrote that gender "refers to the social categories of male and female". This definition aligns with the notion that gender is not solely determined by biology but encompasses the roles and expectations assigned to individuals based on their perceived gender identity.

Both the APA's perspective & Helgeson's definition indicate that gender is multifaceted and influenced by cultural, societal, and psychological factors, going beyond the biological dimension. This approach to gender recognizes the complexity of human identity and how it is constructed within various social contexts.

Field Practical Assignment for Students

Assignment 1

When veterinary college students visit dairy and poultry farmers, they might observe that specific tasks and responsibilities are often associated with gender. For instance, they might notice that milking cows is traditionally seen as a female task due to care given by women folk to animals, while milk marketing is considered a male task. This illustrates how societal expectations and roles associated with gender can shape the division of labour on farms, even if the biological differences between male and female workers are not the sole determinants of these roles.

Assignment 2

Similarly, they may observe that within dairy farming, men are often seen as the primary decision-makers, while women may not be involved in decision making aspects of dairy farming although they play active role in performing various tasks related to dairy farming In contrast, in poultry farming, the roles might differ, with women having more involvement in decision making in managing the flocks. This illustrates how cultural and social norms influence the roles and expectations within these social groups, even within the same animal husbandry sector. Variations also could be noticed not only in different farming systems but also within the farming system depends upon several factors.

II. Gender: Traits and characteristics resulting from social origins

i. In line with the observations of Goldberg's (2010), gender serves as a representation of social and cultural influence on males and females. consider a rural livestock farming society as an illustrative example. Within such a community, specific tasks associated with animal husbandry, such as herding cattle or milking cows, are often closely linked to gender roles. It is common for men to be primarily responsible for herding cattle, while women may take charge of milking cows. These gender roles are not solely determined by biology but are predominantly shaped by the prevailing social and cultural influences within the community. Furthermore, these roles often extend beyond mere labour allocation and can significantly impact access and control over resources and decision-making authority within the context of livestock management.

ii. Denmark *et al.* (2005) defined gender as "a social construction that refers to how differences between girls and boys and women and men are created and explained by society".

Gender: Traits and characteristics shaped by social origins

i. **Toy Preferences**: In societies deeply rooted in traditional gender roles, boys are often encouraged to engage with toys like action figures and building blocks, reinforcing values of strength and problem-solving. Conversely, girls are typically guided towards dolls and domestic playsets, emphasizing qualities associated with nurturing and caregiving.

ii. **Clothing Norms**: Cultural norms surrounding attire often dictate that boys wear pants and shirts, while girls don dresses and skirts. These norms not only reflect societal expectations of appearance but also contribute to the perpetuation of traditional gender roles.

iii. **Career Choices**: Societal expectations and stereotypes frequently steer males towards careers in fields such as engineering, technology, and finance, while females are more often directed towards professions like nursing, teaching, or social work. These career choices mirror cultural perceptions of gender-linked strengths and abilities.

iv. **Emotional Expression**: Boys are often socialized to conceal emotions like sadness or fear, as these may be perceived as signs of vulnerability. In contrast, girls are encouraged to express their emotions openly, reinforcing the stereotype that females are more empathetic and emotionally connected.

v. **Sports Participation**: Cultural norms may lead boys to engage in more aggressive and competitive sports, while girls are directed towards activities perceived as graceful or non-contact, such as ballet or gymnastics. These choices can reflect societal beliefs about gender-associated physical capabilities.

vi. **Household Chores**: Girls are commonly taught to shoulder responsibilities related to cooking, cleaning, and caregiving, reinforcing the notion that these tasks are inherently feminine. Boys, on the other hand, may receive less encouragement to participate in these activities, further entrenching the idea that they belong to the female sphere.

vii. **Media Representations**: In children's media, male characters are often depicted as adventurous and strong, while female characters are portrayed as in need of rescue or as supportive sidekicks. These portrayals not only reflect but also perpetuate traditional gender roles and expectations.

viii. **Parental Roles:** Mothers are frequently presumed to be the primary caregivers for children, while fathers are expected to be the primary breadwinners. These roles reflect historical gender norms and can significantly influence how household responsibilities are distributed.

ix. **Language Use**: Certain words or phrases may be gender-associated. For example, terms like "bossy" might be more commonly used to describe assertive females, while males in similar roles might be seen as displaying leadership qualities.

x. **Social Interactions**: Boys may often be discouraged from displaying vulnerability or seeking emotional support from their peers, reinforcing ideals of independence and self-reliance. In contrast, girls may be encouraged to form close-knit social circles and express emotions openly.

In livestock production, decision-making processes related to the sale or trade of livestock may exhibit distinct gender dynamics. This can manifest as a traditional preference for men holding more influence and authority in such decisions due to deeply ingrained social and cultural norms. Importantly, these gender dynamics may vary from one rural livestock farming society to another, reflecting the diversity of cultural and social contexts.

III. Gender: Categories related to traits considered to be feminine or masculine

i. Wood (1999) defined gender in terms of feminine or masculine qualities, saying that gender refers to how much of each an individual has and to how individuals see themselves: Each of us has some qualities that our culture labels feminine and some it defines as masculine. How much of each set we have indicates our gender. Gender refers to how an individual sees himself or herself in terms of masculine or feminine tendencies

ii. Rider (2005) defined gender as "psychological traits of masculinity and femininity that develop through socialization" and as "the masculine or feminine behaviours that develop through socialization.

Example in a rural livestock farming society

i. **Feminine and Masculine Qualities:** Imagine a society where traditional gender roles are reinforced in livestock farming. Tasks like milking cows, nurturing young animals, and tending to the health and well-being of the livestock might be considered "feminine" qualities. On the other hand, tasks like plowing fields, building fences, taking the animals to markets and restraining large animals might be seen as "masculine" qualities.

ii. **Gender Identity Based on Qualities:** Let's consider a young woman named Geeta in this rural farming community. She excels at caring for the animals, ensuring their nutrition, and tending to their health. Geeta may see herself as having a strong feminine identity because she embodies qualities that her culture associates with femininity. She identifies with her role as a caregiver to the livestock.

iii. **Gender Perception in Farming Roles:** In the same society, a young man named Gopal might take pride in his role as a wood cutter, fence builder, loading and unloading of his bullock cart. He sees himself as having predominantly "masculine" qualities because he excels in tasks typically associated with men in the community. He identifies with his role as a provider and protector of the farm.

iv. **Society's Definition of Gender:** The rural farming society's definition of gender is based on how individuals align with and embody qualities that are culturally labeled as masculine or feminine. It's not just about biological sex but also about the roles and behaviors individuals adopt within the context of farming.

v. **Variability in Gender Identity:** Not everyone in society conforms strictly to these gendered farming roles. Some individuals may have a mix of qualities seen as both feminine and masculine. For instance, a young woman Tara may excel at plowing fields, challenging traditional gender expectations, and causing her community to reconsider their notions of gender.

In this rural livestock farming society, gender identity is not solely determined by one's biological sex but is influenced by how individuals perceive themselves and their alignment with qualities traditionally associated with femininity or masculinity within the context of farming. Gender, in this sense, is a complex interplay of cultural expectations, personal identification, and the roles individuals take on in their agricultural practices.

Gender: Categories related to traits considered to be feminine or masculine

Hypothetical Example

For example, consider a group of college students who are working with dairy farmers. In their interactions, they might observe that certain tasks like milking cows or feeding the animals are often associated with traditionally masculine qualities like physical strength and assertiveness. Meanwhile, tasks related to caring for the animals or managing the farm may be associated with traditionally feminine qualities such as nurturing and attention to detail.

The students might also notice that in the poultry farming sector, tasks such as managing the incubation process or monitoring the health of the birds are often associated with attention to detail and nurturing, the traits are often labelled as feminine. In contrast, tasks such as heavy lifting or equipment maintenance may be linked to traits like physical strength, considered masculine.

College students would understand that these gender roles and behaviours are not inherent but are shaped by the norms and expectations within the farming community. Farmers may adopt or reject certain behaviours based on societal influences and how they have been socialized within the industry in addition to their farming situation. This could lead to variations in how different farmers express their gender through their roles and behaviours on the farm.

In both definitions, college students working with dairy and poultry farmers would gain insight into how gender is a complex interplay of qualities, behaviours, and roles influenced by societal and cultural factors, extending beyond biological distinctions.

IV. Gender: Stereotypes or Expectations that Society Attributes to Women and Men

i. Unger 1993 defined gender as "those characteristics and traits the society considered appropriate to males and females". Unger's definition of gender as can be illustrated in the context of a rural livestock farming society. In such a society, certain traits and behaviours may be deemed appropriate based on one's gender: Example: In this setting, it might be considered appropriate for men to be assertive and independent when making decisions about herd management and finances. In contrast, women may be expected to exhibit nurturing qualities when caring for animals or tending to household responsibilities.

ii. According to Denmark *et al.* (2005), gender comprises traits, interests, and behaviours ascribed to each sex by societies. This concept can be exemplified in a rural farming community: Boys in this community might be encouraged to engage in physically demanding tasks like herding cattle or repairing fences, shaping their interests and behaviours. Girls, on the other hand, may be directed towards activities such as gathering eggs, feeding animals, or assisting with smaller livestock management tasks, influenced by gendered expectations.

iii. Lips (2008) defined gender as the system of societal expectations regarding feminine and masculine roles.

In a rural livestock farming society, these expectations manifest in specific gender roles. Example: society may expect men to primarily serve as the primary breadwinners by selling dairy products, while women are expected to manage the household, including tasks like food preparation and childcare. These gender roles are reinforced within the community.

iv. Etaugh and Bridges (2010) defined gender as "the meanings that societies and individuals give to female and male categories." This definition underscores the influence of cultural and societal norms on gender.

Example: In this society, the meaning of being male might be associated with strength and responsibility for tasks like ploughing fields or constructing animal shelters. Conversely, being female might be associated with nurturing and tending to young animals or maintaining the cleanliness of the homestead. These meanings are constructed by societal norms and expectations.

v. Rothenberg (2004) defined gender as the "socially constructed meanings associated with each sex." This definition emphasizes the social construction of gender roles and expectations.

Example: A young boy growing up in this society may be taught from a young age to aspire to become a skilled farmer and leader, aligning with the socially constructed expectations of masculinity. In contrast, a young girl may be encouraged to learn domestic skills and embrace her role as a caregiver, aligning with the socially constructed expectations of femininity. This highlights that gender meanings are not fixed but shaped by societal norms and beliefs.

In summary, all these definitions converge on the idea that gender is a social construct, moulded by cultural and societal expectations and beliefs. They acknowledge that gender is not solely determined by biology but is rather a complex interplay between individual identities and societal norms.

V. Definitions of Gender by IFAD & European Institute for Gender Equality

IFAD (2010) referred Gender to culturally based expectations of the roles and behaviour of women and men.Unlike the biology of sex, gender roles, behaviours and the relations between women and men are dynamic. They can change over time and vary widely within and across a culture, even if aspects of these roles originated in the biological differences between the sexes.

More comprehensive definitions of sex and gender were provided by The European Institute for Gender Equality, an autonomous body of the European Union

"Sex refers to the biological and physiological characteristics that define humans as female or male. These sets of biological characteristics are not mutually exclusive, as there are individuals who possess both, but these characteristics tend to differentiate humans as females or males."

"Gender refers to the social attributes and opportunities associated with being female and male and to the relationships between women and men and girls and boys, as well as to the relations between women and those between men. These attributes, opportunities and relationships are socially constructed and are learned through socialisation processes. They are context- and time-specific, and changeable. Gender determines what is expected, allowed and valued in a woman or a man in each context. In most societies, there are differences and inequalities between women and men in responsibilities assigned, activities undertaken, access to and control over resources, as well as decision-making opportunities. Gender is part of the broader sociocultural context. Other important criteria for sociocultural analysis include class, race, poverty level, ethnic group and age.

Example in a Livestock rearing community

Livestock rearing involves several activities which are usually shared by the members of the families involved in it. Gender roles and the socially constructed attributes associated with being female and male can be observed:

Role of women and men

Within the livestock rearing family women are primarily responsible for cleaning the animal sheds, collecting the grass, feeding, watering and milking the animals. Alongside their animal care duties, women are responsible for household management, including cooking, cleaning, and childcare. Whereas men are associated with activities which involve more physically demanding tasks such as building and repairing animal enclosures, herding animals to pasture, and veterinary care.

Socially Constructed Attributes in this Context

In this livestock rearing community, the socially constructed attributes associated with being female include qualities related to caregiving, household management, and support roles in animal care.

The socially constructed attributes associated with being male include qualities related to physical strength, leadership in livestock management, and decision-making power in the livestock business.

These socially constructed attributes shape not only the division of labour but also expectations for behaviour, authority, and access to and control over resources within the community.

For example, a young boy growing up in this community may be taught that his role is to become skilled in physically demanding tasks related to livestock management, reflecting the socially constructed expectations of masculinity. A young girl, in contrast, may be encouraged to develop caregiving skills and manage household responsibilities, aligning with the socially constructed expectations of femininity.

This example demonstrates how gender, as defined in the context of the livestock rearing community, involves socially constructed attributes, roles, and expectations. These roles are not inherent to biological sex but are shaped by cultural norms, values, and traditions within the specific community. Just like in human societies, gender roles in this context can vary and are subject to change, highlighting the context- and time-specific nature of gender.

Gender Concept: Story of Rahul and Pooja

Imagine a fictional society called "Mohanpur." In Mohanpur, gender roles and expectations are deeply ingrained in their culture. According to societal norms, women are expected to be caregivers, nurturing, and taking care of the household and children. Men, on the other hand, are expected to be providers, responsible for earning income and making important decisions for the family.

Scenario: Career Aspirations in Mohanpur

Meet Pooja and Rahul, both 18 years old and childhood friends in Mohanpur. Pooja has always been interested in art and dreams of becoming a successful painter, while Rahul is passionate about science and aspires to become a renowned physicist.

i. **Gender Expectations**: Pooja faces societal pressure from family and friends to pursue a career that aligns with their assigned female gender. Painting is considered a hobby, but not a serious career path for women. Meanwhile, Rahul's interest in physics is encouraged, as it aligns with the expectations for their assigned male gender.

ii. **Access to Resources**: Rahul has access to more resources for his career pursuit. He is provided with advanced science education and encouraged to attend workshops and conferences in the field. On the other hand, Pooja 's access to art supplies and formal training is limited due to the perception that her artistic endeavours are more of a pastime.

iii. **Decision-Making Opportunities:** As they grow older, Rahul is given more autonomy in making decisions about their education and future career path. His opinions and aspirations are taken seriously by his family and mentors. However, Pooja faces resistance when expressing her desire to attend an art school, as it is seen as an unconventional choice for someone assigned female.

iv. **Sociocultural Dynamics**: Mohanpur's culture reinforces these gender roles. Media, advertisements, and even educational materials portray women in traditional caregiving roles and men in positions of power and authority. These representations further shape the aspirations and self-perception of individuals like Pooja and Rahul.

In this hypothetical scenario, Mohanpur's societal norms and expectations create disparities in opportunities, access to resources, and decision-making authority based on gender. These inequalities are socially constructed and learned through the process of socialization. Gender in combination with other sociocultural factors such as class, race, and age influence individuals' experiences and life choices.

Differences between Sex and Gender

1. The term "sex" should be used as a classification according to the reproductive organs and functions that derive from the chromosomal complement. In other words, sex refers to the biological differences between males and females, such as differences in reproductive organs, hormones, and chromosomes.

   In contrast, the term "gender" should be used to refer to a person's self-representation as male or female, or how that person is responded to by social institutions based on the individual's gender presentation. Gender refers to the social and cultural differences between males and females, such as differences in gender roles, behaviours, and expectations. While sex is typically thought of as a binary (male or female), gender is more complex and can encompass a range of identities beyond the traditional male/female binary.

2. Sex refers to the biological characteristics that differentiate males from females. These include factors such as gonads (testes or ovaries), sexual

organs (penis or vagina), chromosomes (XX for females and XY for males), and hormonal profiles (oestrogen and progesterone in females, testosterone in males). Sex is typically assigned at birth based on these biological attributes.

Gender is a multifaceted concept influenced by social, psychological, and cultural factors. It is not inherently tied to one's biology but is developed through the process of socialization. Different societies and cultures have varying understandings of what is considered 'masculine' or 'feminine.' Gender roles, norms, and expectations can differ significantly across regions and time periods.

For example, in many rural farming societies, specific gender roles are assigned in the context of animal husbandry. In such societies, men often take on roles perceived as 'masculine,' such as managing and tending livestock, ploughing fields, building and repairing infrastructure like barns and fences, and handling physically demanding tasks. These activities are crucial for the economic sustainability of the family or community.

In contrast, women in these societies may be assigned roles that are considered 'feminine,' such as milking cows, collecting grasses, feeding smaller animals, and converting milk into products such as curd and ghee. They may also be responsible for tasks related to vegetable gardens and household chores, including cooking, cleaning, and childcare. These roles align with cultural expectations and traditions, but it's essential to note that gender roles are not permanent and can change over time and across different cultures.

In summary, while sex is rooted in biological differences, gender is a socially constructed concept shaped by psychological and cultural factors, leading to the assignment of roles and expectations within a given society or culture. These gender roles can vary significantly across different communities and time periods.

3. Sex is typically assigned at birth in most societies, based on observable biological characteristics such as genitalia. However, it's important to note that societies construct norms and expectations related to gender, which individuals learn throughout their lives. These societal influences are pervasive, shaping behaviours and roles for everyone within the community.

   In rural farming societies, gender roles are often pronounced, particularly in activities related to animal husbandry. Men are typically assigned roles that are considered emblematic of strength and responsibility.

Conversely, women in such societies are typically tasked with roles seen as nurturing and domestic. These gendered roles in livestock farming are instilled from a young age as children observe and participate in family farming activities. Parents and grandparents play a significant role in passing down these roles and expectations. Additionally, these gendered roles are reinforced through socialization within the community.

While media influences may be limited in rural areas, local traditions, oral storytelling, and community norms play a significant role in perpetuating these gendered expectations. Over time, these roles become deeply ingrained, affecting the division of labour and shaping the identity and status of individuals within the community.

4. Sex and gender are two distinct concepts that often overlap but are not synonymous. Sex typically refers to the biological characteristics associated with being male or female. However, it's crucial to recognize that sex is not entirely binary, as intersex individuals are born with variations in biological traits.

   On the other hand, gender pertains to a person's internal sense of identity and how he perceives himself in terms of being a man, woman, both, neither, nor anywhere along the gender spectrum. Gender identity is deeply personal and may or may not align with an individual's biological sex. Transgender individuals, for instance, are those whose gender identity differs from their assigned sex at birth.

   Gender norms are societal expectations regarding how individuals should behave and present themselves based on their perceived gender. These norms often revolve around traditional notions of masculinity and femininity and are, unfortunately, rooted in a heteronormative framework, assuming that there are only two genders, male and female, and that these genders are attracted to each other. However, it's important to acknowledge that gender norms are not universal and can evolve over time. Many societies have deeply ingrained traditional gender roles, where men are expected to perform physically demanding tasks, while women are tasked with domestic responsibilities. This binary and heteronormative perspective can lead to the exclusion, discrimination, and violence faced by those who do not conform to these roles.

Difference between Gender & Sex

Aspect	Sex	Gender
	Biological characteristics such as chromosomes, hormones, and reproductive organs that classify individuals as male, female, or Intersex. People are born with it	Socially constructed roles, behaviors, & Expectations(created by people around us) associated with being male, female, or other gender identities. It is learned (parents, siblings teach us these roles)
Changeability	Generally remains unchanged throughout life without significant intervention.	Can be influenced, learned, and changed over time through socialization, education, and personal experiences.
Consistency	Consistent across different cultures and historical periods.	Varies across societies, cultures, and historical periods.
Appearance	Physically recognizable traits typically associated with being male, female, or intersex.	Behavioral, societal, and psychological characteristics associated with being male, female, or other gender identities.
Examples	Male, Female, Intersex	Masculine, Feminine, Non-binary, Genderqueer, Genderfluid, etc.

Gender Equality Versus Gender Equity

Gender equality means equal rights, responsibilities and opportunities of women and men and girls and boys. Gender equality means that all individuals, regardless of gender, have the freedom to pursue personal development and make choices without being restricted by stereotypes or societal norms. It involves recognizing, valuing, and promoting the diverse behaviours, aspirations, and needs of both women and men equally. Importantly, gender equality doesn't imply that men and women should be identical but rather that their rights, responsibilities, and opportunities shouldn't be determined by their gender at birth.

Gender equity is provision of fairness and justice in the distribution of benefits and responsibilities between women and men. It is the process of being fair to women and men. To ensure fairness, strategies and measures must often be available to compensate for women's historical and social disadvantages that prevent women and men from otherwise operating on a level playing field. Equity leads to equalityGender equity focuses on treating women and men fairly based on their specific needs. This may involve equal treatment or different treatment, but it should be regarded as equivalent in terms of rights, benefits, obligations, and opportunities.

Gender and Livestock production

Gender in Livestock production- Definition

Gender in animal husbandry refers to the roles, responsibilities, and social expectations associated with individuals based on their perceived or self-identified gender identity. It encompasses the ways in which people of different genders participate in various aspects of animal farming, including decision-making, labour, and the management of livestock.

Gender Relations in Livestock Production

Gender relations are crucial in livestock production and involve the systematic differentiation of men and women in processes of production. These relations encompass economic, social, and technological interactions necessary for the survival, production, and reproduction of livestock-based livelihoods.

i. **Power Dynamics**: The gender relations in livestock production often reflect the distribution of power between the sexes. Gender is used as a category of social differentiation to delineate differences between genders and may sustain and reinforce inequalities between men and women.

ii. **Changeable Nature of Gender**: Because gender is shaped by learned behaviours and cultural norms, it can change over time and across different contexts. It's not a unilateral concept focused solely on women but is a dynamic relationship between men and women.

iii. **Importance in Livestock Interventions**: Understanding the nature of gender dynamics is essential for ensuring that livestock-based interventions can improve the welfare of both men and women and avoid unintended consequences, particularly in rural settings.

iv. **Key Categories in Gender Dimensions**: Several important categories should be considered when exploring gender dimensions in livestock-based farming systems. These include:

   a. Access to and control over resources.

   b. Division of labour in livestock management.

   c. Access to livestock technologies.

   d. Access to capacity building programmes in livestock production.

   e. Contribution to household income

v. **Gender Imbalances**: Gender imbalances exist across the farming systems and it's crucial to understand the extent to which these imbalances either hinder or support livestock development efforts.

Gender equality and gender equity in the context of livestock production

Gender equity means ensuring fairness and justice for both women and men. It acknowledges that historically and socially, women have faced disadvantages that have prevented them from having the same opportunities as men. To achieve equity, we may need to provide extra support or resources to compensate for these disadvantages. For example, offering scholarships specifically for girls in STEM fields to address the underrepresentation of women in these areas.

Gender equality, on the other hand, means that women and men have equal access to opportunities, resources, and rewards that society values. It's about creating a level playing field where both genders can thrive without facing discrimination or barriers based on their sex. For instance, ensuring that both men and women have the same opportunities for leadership roles in organizations and equal pay for equal work.In simpler terms, gender equity focuses on addressing existing imbalances and providing fair treatment, while gender equality aims for a state where everyone has the same opportunities and rights regardless of their gender.

Dairy Farming

Gender Equity

In many dairy farming communities, men might typically manage the financial aspects of the operation, while women handle tasks like milking and processing dairy products. To promote equity, training programs can be provided to both men and women in financial management, marketing strategies, and technology adoption. For instance, providing women with training on business management software or marketing techniques can help them participate more actively in decision-making processes related to the dairy farm.

Gender Equality

Gender equality in dairy farming ensures that both men and women have equal access to profits generated from the dairy operation. It means that women can also have ownership rights over the dairy animals and have equal say in how the income generated from dairy products is utilized for the benefit of the family.

Poultry Farming

Gender Equity: In poultry farming, tasks like feeding, cleaning, and egg collection may be traditionally assigned to women, while men handle construction of coops or marketing of poultry products. Efforts to promote gender equity may involve providing women with training in poultry management techniques, access to information about market trends, and opportunities to participate in decision-making processes related to the poultry farm.

Gender Equality: Gender equality in poultry farming ensures that both men and women have equal access to resources such as high-quality feed, vaccines, and technical support. It means that women can also access extension services, attend training workshops, and receive credit for investing in poultry farming ventures.

Rural Regions

Gender Equity: In rural regions, women often face challenges accessing extension services, credit facilities, or markets for their produce. Efforts to promote equity might involve setting up women-friendly extension services, establishing women's cooperatives for collective marketing of agricultural products, and providing access to microfinance institutions that offer loans specifically tailored to women's needs.

Gender Equality: Gender equality in rural regions ensures that both men and women have equal opportunities to participate in community decision-making processes related to agriculture,livestock and rural development. It means that women have equal access to education, healthcare, and infrastructure facilities such as roads and irrigation systems that are essential for agricultural,livestock productivity and livelihood improvement.

Summary in Bullet Points

- "Sex" is construed as biological and physiological characteristics including reproductive organs, chromosomes, and hormones that differentiate humans as male or female
- Biological sex is typically categorized as male or female, but variations and intersex individuals exist.
- "Gender" is described as a social construct encompassing roles, behaviours, activities, and attributes that a society considers appropriate for individuals based on their perceived or self-identified gender identity.
- Gender is shaped by societal expectations, norms, and cultural influences, going beyond biological sex.

- Gender influences roles, behaviours, and individual characteristics, and it is learned through socialization.
- It is important to understand the impact of gender on tasks and responsibilities within social groups, even in the absence of significant biological differences.
- It explores the stereotypes and expectations attributed to individuals based on their gender and how these influence behaviour and societal norms.
- Gender intersects with other sociocultural factors like class, race, poverty, ethnicity, and age, shaping individuals' experiences and opportunities.
- It is of paramount importance to understand the nature of gender dynamics, especially in livestock production, to improve the welfare of both men and women and avoid unintended consequences.
- Gender imbalances exist in various aspects of livestock production, and addressing these imbalances is essential for achieving more equitable and inclusive outcomes.
- Culture, society, and history have profound influence on perceptions of gender and sex in contemporary society.

SECTION B : DESCRIPTIVE QUESTIONS

1. Why it is important to study difference between Gender and Sex?

It is important to study the difference between gender and sex because it helps us to understand the complex interplay between biology and culture in shaping human behaviour and development. Recognizing the distinction between sex and gender allows us to move beyond simplistic biological determinism and to appreciate the ways in which social and cultural factors can influence individual differences and group-level patterns. This, in turn, can help us to develop more nuanced and effective interventions to address issues related to gender inequality, discrimination, and bias.

Additionally, understanding the difference between sex and gender can help us to better understand the experiences of individuals who do not conform to traditional gender norms or who identify as transgender or non-binary.

Understanding the concept also is important to promote tolerance and reduce discrimination based on sex, gender, and sexual orientation, which can help create a more inclusive and welcoming learning environment.

2. Why is it important to distinguish between biological aspects of sex, such as chromosomes and reproductive anatomy, and other aspects of gender identity?

It is essential to distinguish between biological aspects of sex and gender identity because they are not always aligned. Gender identity is a deeply personal and psychological concept, whereas biological sex is based on observable physical characteristics. Understanding this distinction helps us respect and acknowledge individuals whose gender identity may not align with their biological sex, promoting inclusivity and respect for diversity.

3. Why is it important to understand the concepts of Sex and Gender for livestock development practitioners?

It is important to understand the concepts of sex and gender for livestock development practitioners for the following reasons

1. Inclusive Development: Livestock Development initiatives must be inclusive and address the needs and rights of all individuals, regardless of their sex or gender identity. Failure to do so can result in marginalized groups being left behind, perpetuating inequalities. For example, in many agricultural and pastoralist communities, livestock farming is a primary source of income and livelihood. However, traditional gender roles and stereotypes often lead to unequal access to and control over livestock resources.
2. Gender-Based Discrimination: Discrimination based on gender is a pervasive issue worldwide. Livestock development practitioners need to understand these dynamics to design interventions that combat gender-based discrimination and promote gender equality. Social exclusion based on gender may happen due to lack of understanding of these concepts as marginalised gender may not be considered when designing and implementing development programs.
3. Economic Empowerment: Gender inequalities in the livestock workforce can hinder livestock based economic development. For example, female livestock farmers may struggle to access loans or modern farming equipment, making it harder for them to improve their livestock production and overall income. Understanding the role of gender in economic participation, including wage gaps and access to employment opportunities, is essential for creating strategies that promote economic empowerment for all.
4. Cultural Sensitivity: Cultural norms and practices related to sex and gender can vary widely across regions and communities. Livestock

development practitioners must be culturally sensitive and aware of these variations to engage effectively with local populations. In certain communities, livestock ownership, particularly among women, can be a source of economic empowerment and independence. Livestock development initiatives should recognize and support these opportunities for women's economic advancement.

5. Effective community engagement: Rural development practitioners work closely with communities to identify their needs and develop sustainable solutions. Understanding sex and its biological aspects allows practitioners to engage with community members more effectively, respecting cultural norms and sensitivities while addressing issues related to health, gender, and livelihoods.

4. How can researchers ensure they are using the terms "sex" and "gender" appropriately in their studies?

Torgrimson and Minso (2005) proposed three guidelines for using sex and gender correctly in human and animal research. First, in the study of human subjects, the term sex should be used as a classification according to the reproductive organs and functions that derive from the chromosomal complement. Second, in the study of human subjects, the term gender should be used to refer to a person's self-representation as male or female, or how that person is responded to by social institutions based on the individual's gender presentation. Third, in most studies of nonhuman animals, the term sex should be used. These guidelines are based on a standard set of definitions for use in sex-based research and can help ensure that researchers are using the terms sex and gender appropriately in their studies. In summary, it is appropriate to use the term sex when referring to the biology of human and animal subjects, and the term gender is reserved for reference to the self-identity and/or social representation of an individual.

5. What is the history of distinction between Gender & Sex?

The distinction between sex and gender began with John Money and his colleagues in the 1950s. They used the term "gender" to refer to the social and cultural aspects of being male or female, while "sex" referred to biological differences. However, the distinction between these terms was not widely adopted until the 1970s, when feminist psychologists began to use the term "gender" to highlight the ways in which social and cultural factors shape gender roles and identities. Since then, there has been ongoing debate and discussion about the meaning and use of these terms, with some authors taking an essentialist stance and others adopting a social constructionist perspective. Despite these differences, most researchers agree that it is important to distinguish between sex and gender to better understand the complex interplay between biology and culture in shaping human behaviour and development

6. What are the implications of distinguishing between Sex and Gender?

Distinguishing between sex and gender has several implications. First, it allows us to recognize that biological sex and gender identity are not always congruent, and that individuals may identify with a gender that is different from their biological sex. This recognition is important for understanding the experiences of transgender and non-binary individuals and for developing interventions to address issues related to gender identity and expression. Second, it allows us to move beyond simplistic biological determinism and to appreciate the ways in which social and cultural factors can influence individual differences and group-level patterns. This, in turn, can help us to develop more nuanced and effective interventions to address issues related to gender inequality, discrimination, and bias. Finally, distinguishing between sex and gender allows us to study the social and cultural aspects of gender in a more systematic and rigorous way, which can help us to better understand the complex interplay between biology and culture in shaping human behaviour and development.

7. What are Heteronormative Expectations?

In certain communities, there is a strong expectation that individuals of different genders will enter heterosexual marriages to raise families and uphold traditional practices. This expectation assumes of a binary gender system and the idea that individuals of different genders are naturally attracted to each other. In such communities, individuals who do not conform to these binary gender roles or who identify as LGBTQ(Lesbian, Gay, Bisexual, Transgender and Queer) + may experience severe exclusion and discrimination. For example, someone assigned male at birth who identifies as female or non-binary may face ostracism for expressing his true gender identity and not adhering to traditional male roles. In extreme cases, LGBTQ+ individuals may encounter violence, harassment, or social isolation due to their failure to conform to expected gender and sexual norms. This mistreatment can manifest as physical violence, verbal abuse, or social alienation, causing significant harm and distress. Those who do not fit within traditional gender roles may also find themselves with limited access to resources, opportunities, or leadership positions within their community. For instance, a non-binary individual might struggle to gain acceptance as a community leader in decision-making processes. Gender is a highly individual and complex aspect of human identity. Some people recognize their gender identity early in childhood, while others may do so later in life. Gender is not solely defined by biological sex; it is about how individuals perceive themselves and their identity. Furthermore, gender does not exist in isolation; it intersects with other aspects of an individual's identity, such as class, race, ethnicity, religion, and disability. Gender expression encompasses a wide range of behaviours and characteristics, including how a person dresses, moves, styles his hair, and interacts with others. In summary, acknowledging the distinction between sex and gender and challenging traditional gender norms is essential for promoting inclusivity, diversity, and respect for all individuals, regardless of their gender identity or sexual orientation.

8. How is the concept of gender as a social construct relevant in working with livestock farmers for veterinary/livestock practioners?

The veterinary/livestock practioners during their working with livestock farmers in a rural community may encounter gender-related factors in the following ways:

Roles in Animal Care: In some cultures, there may be traditional gender roles in animal care. For instance, men might be more involved in tasks like herding and feeding, while women could be responsible for milking and cleaning the barn. Understanding these roles can help provide tailored advice and support to farmers, considering the gender dynamics at play.

Decision-Making: Gender norms can influence decision-making in livestock management. In some cases, male farmers might make the major decisions regarding animal health and breeding, while female farmers play a key role in managing household resources. So one should be aware of these dynamics and ensure that their recommendations and information reach all decision-makers, regardless of gender.

Access to Resources: The availability and access to veterinary services, financial resources, and training can differ based on gender. For instance, women in farming may face barriers in accessing information on technologies or loans for livestock improvement. So practioners can ensure that their services are accessible and beneficial to all members of the community.

Awareness and Education: Understanding that gender plays a role in knowledge dissemination is essential. They may notice that rural women interact freely with female extension agents compared to male extension agents & can use gender-sensitive approaches in their education efforts. They might organize workshops or training sessions that consider the specific needs and schedules of both male and female farmers, ensuring that information reaches everyone effectively.

By considering the socially constructed roles and expectations related to gender, veterinary practioners can be more sensitive to the diverse needs and situation of livestock farmers. This awareness can help them provide better support and guidance to the farmers they work with, ultimately contributing to more inclusive and effective delivery veterinary services in the community.

9. How can the understanding of gender as a social construct benefit veterinarians in their interactions with livestock farmers?

The understanding of gender as a social construct can benefit veterinarians in their interactions with livestock farmers in several ways:

Tailored extension Advisory: Veterinarians can provide more tailored extension advisory to livestock farmers based on their understanding of how societal expectations and norms may influence the roles and responsibilities of male and female farmers. This can help ensure that the advice given is relevant to the specific context of each farm.

Sensitivity to Gender Roles: Veterinarians can be more sensitive to the gender roles and expectations within farming communities. By recognizing these roles, they can avoid making assumptions about who performs certain tasks and instead engage with farmers in a respectful and inclusive manner.

Promoting Equality: Veterinarians can play a role in challenging and promoting equality within farming communities. They can advocate for gender-inclusive practices and encourage open discussions about how traditional gender roles may affect farm management and animal care.

Understanding Decision-Making: Recognizing the influence of gender on decision-making in farming households can help veterinarians understand the dynamics of these households better. They can adapt their recommendations to consider the input and roles of all family members, regardless of gender.

Cultural Sensitivity: Veterinarians can be more culturally sensitive by acknowledging that gender norms and expectations can vary widely across different farming communities. This understanding allows them to engage with farmers in a way that respects and aligns with their cultural context.

Efficient Farm Management: By understanding how gender roles might affect task assignments on the farm, veterinarians can offer advice that promotes efficiency and productivity while respecting the capabilities and preferences of all individuals involved in livestock farming.

Inclusive Practices: Veterinarians can advocate for practices that prioritize animal welfare and productivity over traditional gender-based task assignments. This can help create a more inclusive and efficient farming environment.

In summary, the understanding of gender as a social construct equips veterinarians with the knowledge and sensitivity to navigate the complexities of gender roles and expectations within farming communities. This enables them to provide more effective, culturally sensitive, and inclusive support to livestock farmers.

10. How do veterinarians contribute to creating a more inclusive and productive farming environment by considering the complex interplay between biology and culture in gender roles?

Veterinarians can contribute to creating a more inclusive and productive farming environment by considering the complex interplay between biology and culture in gender roles in several ways:

Educational Initiatives: Veterinarians can organize educational programs and workshops for farmers that highlight the importance of considering both biology and culture in livestock farming. These initiatives can raise awareness about how gender roles may be influenced by tradition and biology.

Task Diversification: Veterinarians can work with farmers to diversify tasks based on individual capabilities rather than traditional gender expectations. By encouraging both men and women to participate in a variety of farm tasks, productivity can increase, and the workload can be shared more equitably.

Advocating for Gender Equality: Veterinarians can advocate for gender equality in livestock farming practices. They can promote practices that prioritize efficiency and animal welfare over gender-based assignments. This can lead to a more balanced distribution of responsibilities and improved productivity.

Inclusive Language: Veterinarians can use inclusive language when communicating with farmers. By avoiding gender-specific language in their recommendations and discussions, veterinarians can help challenge stereotypes and promote inclusivity.

Open Dialogue: Veterinarians can facilitate open dialogues within farming communities about the interplay of biology and culture in gender roles. These discussions can help farmers recognize the value of diversifying roles and responsibilities to enhance farm management.

Respecting Cultural Norms: Veterinarians can respect the cultural norms of each farming community they work with. They can adapt their recommendations to align with local customs and expectations while gently encouraging changes that promote inclusivity and productivity.

Supporting Women in Farming: In many cases, women play significant roles in livestock farming but may not receive due recognition. Veterinarians can actively support and empower female farmers by acknowledging their contributions and ensuring they have access to resources and training. Veterinarians can encourage formation of groups such as women self-help groups, women dairy coops etc to empower them

Advocating for Technology Adoption: Veterinarians can advocate for the adoption of modern farming technologies that can make tasks more accessible and less physically demanding, reducing the influence of traditional gender roles. For example, use of milking machines to reduce the burden on women and time to milk the animals.

Data-Driven Decision-Making: Veterinarians can promote data-driven decision-making in livestock farming. By collecting and analysing data on the impact of diversified roles on farm productivity, they can provide evidence to support more inclusive practices. Veterinarians can help them in maintaining farm records (data on both inputs and outputs) which form the basis for taking appropriate decisions in future.

In summary, veterinarians can contribute to a more inclusive and productive farming environment by considering the interplay between biology and culture in gender roles. By advocating for diversity, inclusivity, and gender equality in farm management, they can help create a more efficient and equitable agricultural landscape.

SECTION C: SHORT ANALYTICAL QUESTIONS

1. **Explain the biological criteria that determine sex and provide an example illustrating how chromosomes, hormones, and reproductive anatomy play a role in defining male and female individuals.**

   In biology, sex is determined by a combination of chromosomes, hormones, and reproductive anatomy. Chromosomes, specifically the presence of XX or XY, play a crucial role in shaping reproductive structures. For instance, individuals with XX chromosomes typically develop female reproductive organs like the uterus and ovaries, while XY chromosomes lead to the development of male reproductive structures such as testes and a penis. Hormones, such as estrogen and testosterone, further differentiate these structures as the fetus grows. An example scenario involves prenatal testing where an expectant couple learns from an ultrasound that their unborn child possesses XY chromosomes, indicating a biological male. This determination is based on the presence of the Y chromosome, leading to the development of male reproductive organs and the production of male hormones.

2. **Discuss the multifaceted nature of gender as a concept in sociology and psychology. Provide definitions from different perspectives, such as the American Psychological Association and Helgeson, and explain how gender extends beyond biological distinctions.**

   Gender, in the realm of sociology and psychology, is a multifaceted concept that transcends simple binary classifications based on biological differences. The American Psychological Association (APA) defines gender as a cultural term, highlighting it when referring to women and men as social groups. Helgeson expands on this by defining gender as social categories of male and female. These perspectives emphasize that gender is not solely determined by biology but is influenced by cultural, social, and psychological factors. The complexity of human identity is recognized through the interplay of societal norms, roles, and expectations in shaping gender.

3. **How do cultural and social norms influence gender roles in animal husbandry settings? Provide practical examples from a hypothetical scenario involving college students working with dairy and poultry farmers.**

   Cultural and social norms significantly impact gender roles in agricultural settings, as observed in a hypothetical scenario with college students working with dairy and poultry farmers. For instance, in dairy farming, tasks like milking may be traditionally associated with femininity due to caregiving roles. In contrast, marketing or heavy lifting may be seen as masculine tasks. Similarly, in poultry farming, roles may differ, with women having more involvement in managing flocks. These examples illustrate how societal expectations shape the division of labour in farming practices, emphasizing the influence of cultural norms on gender roles within the same industry.

4. **Define gender using psychological perspectives, such as Goldberg's observation and Denmark et al.'s definition. Provide examples from a rural livestock farming community to illustrate how gender is socially constructed and influenced by cultural and social factors.**

   Gender, according to Goldberg, represents social and cultural influences on males and females. Denmark *et al.* further define gender as a social construction explaining differences between girls and boys and women and men. In a rural livestock farming community, gender roles extend beyond specific tasks, influencing comprehensive responsibilities and power dynamics. Decision-making processes, access to resources, and authority in areas like livestock trade reflect socially constructed expectations. For example, men may be perceived as primary decision-makers, aligning with traditional norms, while women may be associated with caregiving tasks. These examples demonstrate how gender in this context is shaped by cultural and social influences, illustrating the complex interplay of individual identity and societal norms.

5. **Explore the concept of gender stereotypes and expectations using Lips' and Etaugh and Bridges' definitions. Provide examples from a practical scenario involving veterinary college students working with dairy farmers to illustrate how these stereotypes impact roles and behaviors of men and women in the farming industry.**

   According to Lips, gender involves characteristics and traits sociocultural considers appropriate to males and females. Etaugh and Bridges define gender as the meanings societies attribute to female and male categories. In a practical scenario, veterinary college students working with dairy farmers might observe stereotypes such as associating physically demanding tasks with masculinity and caregiving roles with femininity. These expectations influence the division of labor, decision-making, and societal perceptions within the farming community. For instance, men may be expected to handle physically demanding tasks, while women are associated with caregiving and administrative responsibilities. Challenging and transforming these stereotypes is crucial for promoting gender equality in farming practices.

6. **Define gender according to IFAD and the European Institute for Gender Equality. Provide an example from an animal husbandry rearing community to illustrate how culturally based expectations influence the roles and responsibilities assigned to women and men.**

   IFAD defines gender as culturally based expectations of the roles and behavior of women and men. The European Institute for Gender Equality defines gender as social attributes and opportunities associated with being female and male, shaped by sociocultural context. In an animal husbandry rearing community, roles and responsibilities are culturally influenced. Women may be responsible for daily animal care and household management, aligning with cultural expectations of femininity. Men may take on tasks related to livestock management and income generation, reflecting expectations of masculinity. These culturally based expectations impact access to resources, decision-making, and social status. This example highlights how gender roles in animal husbandry are culturally determined and vary across communities, emphasizing the influence of culture on gender expectations.

7. **According to the American Psychological Association (APA), how is gender defined in the context of sociology?**

   The APA defines gender as a cultural term, referring to women and men as social groups, emphasizing the cultural and societal dimensions beyond biological distinctions.

8. **In a rural livestock farming society, how might specific tasks be associated with gender roles?**
   Tasks like milking cows may be traditionally seen as a female responsibility, while tasks like managing the flock or making decisions related to livestock sales may be associated with males.
9. **How does Goldberg (2010) define gender in terms of social and cultural influences?**
   Goldberg defines gender as representing social and cultural influences on males and females, highlighting how behaviours and traits are shaped by societal expectations.
10. **Give an example of a socially constructed gender role in a rural livestock farming community.**
    Men in the community might be expected to be primary decision-makers in tasks like selling livestock, while women are expected to manage household responsibilities and caregiving tasks.
11. **According to Denmark *et al.* (2005), how is gender defined as a social construct?**
    Gender is defined as a social construction that refers to how differences between girls and boys and women and men are created and explained by society.
12. **How do cultural norms influence gender roles in agriculture, as observed in a poultry farming setting?**
    Cultural norms may dictate that women have more involvement in managing flocks, while men are expected to engage in tasks like heavy lifting or equipment maintenance.
13. **What does Toy Preferences illustrate in the context of traditional gender roles?**
    Toy preferences reinforce traditional gender roles by encouraging boys to play with action figures and girls with dolls, reflecting societal values associated with strength and nurturing, respectively.
14. **How might media representations contribute to the perpetuation of traditional gender roles?**
    Male characters in children's media are often depicted as adventurous and strong, while female characters may be portrayed as in need of rescue or in supportive roles, reinforcing traditional stereotypes.
15. **Define gender identity based on qualities in a rural farming society.**
    Gender identity is how individuals align with qualities culturally labelled as masculine or feminine, influencing their roles and behaviours within the context of farming.

16. **How do societal expectations shape the division of labour in a farming community, as seen in the handling of heavy machinery?**

    Societal expectations may link the handling of heavy machinery with masculinity, influencing the roles and behaviours assigned to men in the community.

17. **According to the European Institute for Gender Equality, how does gender refer to the relationships between women and men?**

    Gender refers to the social attributes and opportunities associated with being female and male and the relationships between women and men, which are socially constructed and learned through socialization processes.

18. **How do culturally based expectations influence roles and responsibilities in animal care within a community?**

    Culturally based expectations shape roles, such as women being responsible for daily animal care, and men managing tasks like breeding and selling animals.

19. **In a rural livestock farming society, how are socially constructed attributes associated with being male reflected in leadership roles?**

    Men may hold leadership positions in community organizations related to animal husbandry, reflecting socially constructed attributes of strength, decision-making, and leadership.

20. **According to Unger's definition, what does gender encompass?**

    Gender encompasses those characteristics and traits sociocultural considered appropriate to males and females.

21. **Give an example of gender stereotypes in a farming community related to decision-making.**

    Men may be expected to be assertive and independent decision-makers on the farm, while women may be expected to exhibit nurturing qualities in caregiving roles.

22. **How does Lips (2008) define gender in terms of societal expectations?**

    Lips defines gender as the system of societal expectations regarding feminine and masculine roles.

23. **What does the term "gender roles" refer to in the context of a rural livestock farming community?**

    Gender roles refer to the socially constructed expectations of tasks and responsibilities assigned to women and men, shaping their roles in animal care and management.

24. According to Etaugh and Bridges (2010), what do societies and individuals give to female and male categories?

Gender is defined as "the meanings that societies and individuals give to female and male categories," highlighting the role of cultural interpretations in shaping gender expectations.

25. What is Goldberg's observation regarding gender?

Goldberg observed that gender represents social and cultural influences on males and females.

26. According to Denmark *et al.*, how is gender defined?

Gender is defined as a social construction that refers to how differences between girls and boys and women and men are created and explained by society.

27. What are some examples of gender stereotypes related to toy preferences?

Boys are often encouraged to play with action figures and building blocks, while girls are directed towards dolls and domestic playsets.

28. How do career choices often reflect gender-related expectations?

Males are frequently directed towards careers in engineering and finance, while females are steered towards nursing, teaching, or social work.

29. According to Wood, how is gender defined in terms of individual qualities?

Wood defines gender as how much feminine or masculine qualities an individual possesses, indicating their gender identity.

30. In a hypothetical livestock farming society, how might traditionally masculine and feminine qualities be associated with specific tasks?

Tasks like herding cattle might be considered masculine, while tasks like milking cows might be associated with feminine qualities.

31. In a rural farming community, what socially constructed attributes might be associated with being male?

Attributes related to physical strength, leadership in livestock management, and decision-making power in the livestock business may be associated with being male.

32. What does Unger's definition emphasize regarding gender?

Unger defines gender as characteristics and traits sociocultural considered appropriate to males and females.

33. According to Rothenberg, how are gender meanings associated with each sex?

Rothenberg defines gender as socially constructed meanings associated with each sex, shaped by societal norms and beliefs.

34. Provide an example of gender expectations in a rural livestock farming society.

In such a society, men might be expected to be assertive and independent decision-makers, while women are expected to exhibit nurturing qualities in caregiving roles.

35. How might gender roles in livestock management impact access to resources and decision-making authority?

Men may have more influence in decisions related to breeding and selling animals, influencing economic outcomes.

36. How does gender, as defined by IFAD, vary across different cultures?

Gender roles and expectations in animal care and management are culturally influenced and can vary significantly from one culture to another.

37. Provide an example of a gender stereotype related to emotional expression ?

Boys are often socialized to conceal emotions like sadness or fear, reinforcing the stereotype that males should avoid displaying vulnerability.

38. How do socially constructed gender roles impact access to resources and decision-making authority in a community?

Socially constructed gender roles can influence access to resources and decision-making authority, with expectations often favoring men in roles associated with leadership and economic decisions.

39. Define "gender as a social construct."

"Gender as a social construct" means that gender is not solely determined by biology but is shaped by societal expectations, norms, and cultural influences. It emphasizes that differences between men and women are influenced by society.

40. What does "gender socialization" refer to?

Gender socialization refers to the process through which individuals learn and internalize societal expectations and roles associated with their perceived gender.

41. How are gender roles, behaviours, and expectations influenced?

Gender roles, behaviors, and expectations are influenced by cultural and societal norms. Society assigns specific roles and characteristics to individuals based on their perceived gender.

42. What is the concept of "gender diversity"?

Gender diversity is the acknowledgment and acceptance of a range of gender identities beyond the traditional binary concept of male and female. It recognizes that gender is not limited to just two categories.

43. Explain the term "gender identity."

Gender identity refers to an individual's personal perception and understanding of their own gender. It may or may not align with their biological sex.

44. What is the role of chromosomes in determining an individual's biological sex?

Chromosomes play a role in differentiating biological sex. Typically, females have XX chromosomes, and males have XY chromosomes.

45. What is the significance of gender diversity in society?

Gender diversity is significant in promoting inclusivity and recognizing the existence of various gender identities beyond the binary. It ensures that individuals of all gender identities are respected and represented.

46. How do societal expectations and norms influence gender roles and behaviours?

Societal expectations and norms shape gender roles and behaviours by assigning specific traits, interests, and roles to individuals based on their perceived gender. These expectations can vary between cultures and over time.

47. What is the definition of gender in the context of animal husbandry?

Gender in animal husbandry refers to the roles, responsibilities, and social expectations associated with individuals based on their perceived or self-identified gender identity. It encompasses the ways in which people of different genders participate in various aspects of animal farming, including decision-making, labour, and the management of livestock.

48. How does gender relate to power dynamics in the context of livestock production?

Gender relations in livestock production often reflect the distribution of power between the sexes. Communities use gender as a category of social differentiation to delineate differences between men and women, which may sustain and reinforce inequalities between them. The power dynamics are influenced by societal norms and expectations related to gender.

49. Why is it important to consider gender dynamics in livestock interventions?

Understanding the nature of gender dynamics is essential for ensuring that livestock-based interventions can improve the welfare of both men and women and avoid unintended consequences, especially in rural settings. Gender dynamics affect how resources are allocated, labor is divided, and technology is accessed, all of which can significantly impact the success of livestock development efforts.

50. How can gender imbalances impact livestock development efforts?

Gender imbalances can hinder livestock development efforts. When there are imbalances in access to resources, division of labor, and decision-making, it can lead to inefficiencies and inequalities in livestock production. Addressing these imbalances is crucial for more equitable and successful livestock interventions.

51. What is the difference between sex and gender in the context of animal husbandry?

In the context of animal husbandry, sex refers to the biological attributes and characteristics that differentiate individuals as male or female based on factors like reproductive organs, chromosomes, and hormones. In contrast, gender refers to the socially constructed roles, responsibilities, and expectations assigned to individuals based on their perceived or self-identified gender identity within the context of livestock production.

52. Why is understanding gender dynamics important in livestock interventions, especially in rural contexts?

Understanding gender dynamics is crucial in livestock interventions, particularly in rural contexts, because it allows for the improvement of the overall welfare of both men and women involved in livestock-based livelihoods. It helps in identifying and addressing gender imbalances that may hinder or support livestock development efforts. Without this understanding, interventions might inadvertently reinforce inequalities and have unintended consequences for different genders.

53. How is gender different from sex in terms of its changeable nature and sociocultural influence?

Gender is different from sex in that it is a socially constructed concept that is shaped and reshaped by cultural norms and societal influences. Gender is changeable over time and across contexts, whereas sex, which is based on biological attributes, is relatively stable. Gender is not a unilateral concept focusing solely on women or men, but rather a dynamic relationship between individuals influenced by sociocultural factors.

54. Why are gender stereotypes and expectations important to consider in the context of livestock production?

Gender stereotypes and expectations are important to consider in the context of livestock production because they influence the roles and behaviours of individuals based on their perceived gender. These expectations may shape the division of labour, access to resources, and decision-making authority in livestock farming. Recognizing and challenging these stereotypes is essential for promoting gender equality in farming practices and ensuring that interventions benefit all individuals, regardless of their gender.

55. How does gender intersect with other sociocultural factors, and why is this intersection important to understand?

Gender intersects with other sociocultural factors such as class, race, ethnicity, religion, and disability. This intersection is important to understand because it influences an individual's experiences and opportunities within society. Different social categories and identities may intersect to create unique challenges and advantages for individuals. Recognizing these intersections is essential for addressing inequalities and promoting inclusivity and equity in livestock interventions and other aspects of life.

56. How do gender relations manifest in livestock production, and why are they important?

Gender relations in livestock production involve the systematic differentiation of men and women in processes of production. They encompass economic, social, and technological interactions necessary for the survival, production, and reproduction of livestock-based livelihoods. Understanding gender relations is vital as they influence power dynamics, inequalities, and the overall success of livestock-based interventions.

57. Why it is significant to recognize that gender is a dynamic concept influenced by cultural norms?

Recognizing that gender is dynamic and shaped by cultural norms is significant because it allows for a better understanding of how roles and expectations can change over time and across different contexts. This awareness is essential for designing effective livestock interventions that consider the ever-evolving nature of gender.

58. How can gender imbalances impact livestock development efforts, and why is it important to address these imbalances?

Gender imbalances in aspects of livestock development can hinder progress and perpetuate inequalities. It's important to address these imbalances to ensure that interventions are equitable and sustainable, benefiting both men and women and avoiding unintended consequences, particularly in rural contexts.

59. How does society use gender as a category of social differentiation in livestock production, and what are the consequences of this differentiation?

Society uses gender as a category to delineate differences in roles and responsibilities within livestock production. This can result in the reinforcement of inequalities between men and women, impacting access to resources, decision-making power, and overall welfare in the context of livestock interventions.

60. **What are the five key areas of consideration when aiming to improve gender outcomes in livestock interventions, and why are they significant?**

The five key areas of consideration are gendered access to and control over livestock, division of labor, access to technologies, access to extension and training, and intra-household dynamics. They are significant because addressing gender imbalances in these areas is essential to promote gender equity and improve the effectiveness of livestock development efforts.

61. **How does the understanding of gender dynamics in livestock production contribute to sustainable development, and why is it crucial in rural contexts?**

Understanding gender dynamics in livestock production is essential for sustainable development because it ensures that interventions benefit both men and women and avoid unintended consequences. This is particularly crucial in rural contexts where traditional gender roles may be more pronounced and where addressing gender imbalances can lead to more equitable and productive livestock systems.

SECTION D: OBJECTIVE QUESTIONS

Question 1: Multiple Choice Questions

1. **What is gender in the context of animal husbandry?**
   a. Biological attributes of individuals
   b. Social roles, responsibilities, and expectations based on perceived gender identity
   c. Chromosomes and hormones
   d. Reproductive anatomy

2. **How is gender different from sex?**
   a. Sex is about biological characteristics; Gender is a social construct.
   b. Gender and sex are identical concepts.
   c. Sex is the result of learned behaviour, while gender is assigned at birth.
   d. Sex is a unilateral concept focusing on women, while gender is dynamic.

3. **Which of the following is an example of a gendered task in animal husbandry?**
   a. Operating irrigation
   b. Handling livestock extension and training
   c. Milking cows
   d. Marketing agricultural products

4. **What does a gender relation in livestock production refer to?**
   a. Biological differences between men and women
   b. Economic, social, and technological interactions in animal farming
   c. Gender identity
   d. Distribution of livestock within the household

5. **How can gender dynamics affect livestock-based interventions?**
   a. They have no impact on interventions.
   b. They can lead to unintended consequences and inequalities.
   c. They primarily benefit men.
   d. They only affect gender-imbalanced societies.

6. **Which of the following is a key category to consider when exploring gender dimensions in livestock-based systems?**
   a. Agricultural policy development
   b. Access to and control over livestock
   c. Weather patterns
   d. Geographic location of farms

7. **In the context of livestock production, what is an example of a gendered division of labour?**
   a. Men and women equally participating in all tasks
   b. Men primarily responsible for cooking
   c. Men handling heavy machinery maintenance
   d. Men and women taking turns in all tasks

8. **Which term refers to the socially constructed meanings associated with each sex?**
   a. Biological sex
   b. Gender identity
   c. Gender roles
   d. Gender stereotypes

9. **What is the definition of "sex" according to the European Institute for Gender Equality?**
   a. It refers to psychological attributes assigned to individuals.
   b. It's a binary classification based on chromosomes and hormones.
   c. It's a socially constructed concept.
   d. It is an individual's self-representation as male or female.

10. How does gender differ from sex according to European Institute for Gender Equality?

a. Gender is a biological concept, while sex is a social construct.

b. Gender is a binary classification, while sex is a fluid concept.

c. Gender is about reproductive anatomy, while sex is about social expectations.

d. Gender is about social roles and behaviours, while sex is about biological attributes.

11. What does the term "gender" refer to according to the American Psychological Association (APA).

a. Biological differences between males and females

b. Social categories distinguished by psychological features

c. A person's self-identification as male or female

d. An individual's physical characteristics

12. How is gender expressed in terms of an individual's self-perception?

a. Through clothing choices, interests, and roles

b. Through chromosomes and hormones

c. Through reproductive anatomy

d. Through social interactions

13. Which factor influences the expression of gender?

a. Biological sex

b. Chromosomes

c. Social and cultural norms

d. Genetics

14. What does gender encompass according to Goldberg (2010)?

a. Biological characteristics that determine sex

b. Psychological and social aspects of being male or female

c. Chromosomes and reproductive anatomy

d. Social and cultural influences on males and females

15. How is gender defined by Denmark et al. (2005)?

a. It refers to the classification of individuals based on their genetic makeup.

b. It represents social and cultural influences on males and females.

c. It is a binary concept with no variation.

d. It is solely based on reproductive anatomy.

16. **How does gender interact with other social categories according to Lips (2008)?**
    a. Gender has no interaction with other categories.
    b. Gender reinforces traditional gender roles.
    c. Gender primarily influences economic factors.
    d. Gender intersects with other social categories like class and race.

17. **What is gender, according to Rider (2005)?**
    a. A biological classification based on chromosomes and hormones
    b. The result of learned behaviours and cultural norms
    c. A fixed concept focusing on women
    d. A unchangeable aspect of an individual's identity

18. **What does gender encompass in the context of animal husbandry?**
    a. Biological attributes of individuals
    b. Social roles, responsibilities, and expectations based on perceived gender identity
    c. Chromosomes and hormones
    d. Reproductive anatomy

19. **What is gender relations in livestock production primarily related to?**
    a. Biological characteristics
    b. Roles assigned based on reproductive anatomy
    c. Economic and social interactions
    d. Genetic differences

20. **How does gender affect power dynamics in livestock production?**
    a. It doesn't have any impact on power dynamics.
    b. It reinforces and sustains power imbalances.
    c. It leads to complete equality between men and women.
    d. It only affects men's roles in livestock production.

21. **What does gender encompass in the context of animal husbandry?**
    a. Biological attributes of individuals
    b. Social roles, responsibilities, and expectations based on perceived gender identity
    c. Chromosomes and hormones
    d. Reproductive anatomy

22. What are gender imbalances often related to in livestock development?

a. Increased efficiency in livestock production
b. Favourable economic conditions for women
c. Hindering the development of livestock-based livelihoods
d. Enhancing access to resources for women

23. In gender relations, what role does power play in livestock production?

a. Power dynamics have no impact on livestock production.
b. Power is equally distributed between men and women.
c. Power dynamics can either reinforce or challenge gender inequalities.
d. Power is solely determined by biological differences.

24. Which term refers to the biological aspects that differentiate males and females, including chromosomes, hormones, and reproductive anatomy?

a. Gender
b. Sex
c. Culture
d. Gender identity

25. According to the European Institute for Gender Equality, how is "gender" defined?

a. Gender is a synonym for sex.
b. Gender is a biological characteristic.
c. Gender encompasses social attributes and roles associated with being male or female.
d. Gender is the same as gender identity.

26. How does gender differ from sex?

a. Gender refers to biological differences, while sex is a social construct.
b. Gender and sex are interchangeable terms.
c. Sex refers to social roles, while gender relates to biology.
d. Gender is a term for male, and sex is a term for female.

27. Which statement is true about gender and sex?

a. Gender and sex are fixed and unchanging concepts.
b. Gender is solely a product of biology.
c. Gender is shaped by cultural norms and can change over time.
d. Sex is determined by psychological traits.

28. How is gender described in terms of traits and characteristics by Wood (1999)?

a. Gender is an innate quality.

b. Gender encompasses physical appearance.

c. Gender is a result of biological attributes.

d. Gender refers to psychological traits of masculinity and femininity.

29. Gender identity can be described as:

a. A fixed and unchangeable aspect of an individual's identity

b. Learned norms and roles imposed by society

c. How an individual feels about themselves in terms of masculinity or femininity

d. Assigned at birth based on physical characteristics

30. What is the term for people who do not identify as strictly male or female?

a. Traditional

b. Heteronormative

c. Binary

d. Non-binary or genderqueer

31. What does sex primarily refer to?

a. Biological & physiological aspects

b. Social roles and expectations

c. Personal identity

d. Cultural norms

32. According to the definitions provided, what determines an individual's biological sex?

a. Chromosomes, hormones, and reproductive anatomy

b. Personal preferences and identity

c. Societal expectations and roles

d. Education and upbringing

33. In the context of gender, what does "gender expression" refer to?

a. Biological characteristics

b. How one feels about themselves

c. How an individual communicates their gender

d. Cultural norms

34. According to Wood's definition, what determines an individual's gender?

a. Biological characteristics

b. Socialization and cultural influences

c. Personal preferences

d. Education and upbringing

35. What is the term for traits and characteristics associated with being male or female?

a. Sex
b. Gender roles
c. Gender stereotypes
d. Gender norms

36. Gender roles and expectations are primarily shaped by:

a. Biology
b. Individual preferences
c. Societal norms and culture
d. Genetics

37. What is the term for the meaning of being male or female, shaped by cultural and societal norms?

a. Biological sex
b. Gender identity
c. Gender expression
d. Gender roles

38. What do traditional gender norms often presuppose?

a. Gender equality
b. Heteronormativity
c. Non-binary identities
d. Gender fluidity

39. What is the primary biological factor that determines an individual's sex?

a. Hormones
b. Chromosomes
c. Genitalia
d. Secondary sexual characteristics

40. In the context of gender, what does "social construct" mean?

a. A biological concept
b. Something that is universally fixed
c. Shaped by societal expectations and norms
d. Determined solely by genetics

41. Which of the following is an example of a gender stereotype?

a. Women typically have two X chromosomes
b. Men are generally taller than women
c. Boys are better at math, and girls are better at language
d. Gender identity is diverse and individual

42. What is gender socialization?

a. A genetic process that determines gender identity
b. The influence of society and culture in shaping gender roles and behaviours
c. A medical treatment for gender dysphoria
d. The biological basis of gender identity

43. Which term refers to the meanings that societies and individuals give to female and male categories?

a. Gender roles
b. Gender identity
c. Gender stereotypes
d. Gender ascription

44. In some cultures, pink is associated with girls and blue with boys. This illustrates the ____________ nature of gender.

a. Biological
b. Inherent
c. Universally fixed
d. Socially constructed

45. How are the psychological features and role attributions of gender typically assigned in society?

a. Based on individual preference
b. Based on biological factors
c. Based on societal expectations
d. Randomly

46. Which of the following is a primary biological difference between males and females?

a. Clothing choices
b. Interests and hobbies
c. Chromosomes
d. Social roles

47. Which hormone is primarily responsible for the development of male secondary sexualcharacteristics, such as facial hair and a deep voice?

a. Estrogen
b. Progesterone
c. Testosterone
d. Prolactin

48. What are the key reproductive organs in females?

a. Testes
b. Uterus and ovaries
c. Penis and scrotum
d. Fallopian tubes

49. What does gender refer to in the context of how individuals see themselves?

a. Social roles
b. Biological sex
c. Gender stereotypes
d. Gender identity

50. In many societies, traits like assertiveness and independence are often associated withwhich gender?

a. Both genders equally
b. Masculinity
c. Femininity
d. Neither gender

51. Gender roles, behaviours, and expectations can vary widely between different societiesand change over time, reflecting the influence of:

a. Genetics
b. Universal standards
c. Cultural and societal norms
d. Biological determinants

52. What is the meaning of gender as a "social construct"?

a. It refers to gender being biologically determined.

b. It signifies that gender is shaped by societal expectations and norms.

c. It suggests that gender is a fixed and inherent quality.

d. It refers to gender as an individual preference.

53. Which characteristic is ascribed to males in many societies?

a. Nurturing and caregiving

b. Physical strength and assertiveness

c. Attention to detail and empathy

d. Traditional female roles

54. Which definition of gender emphasizes that it is not an inherent quality but rather a social construct?

a. Gender is a product of biological factors.

b. Gender is the result of individual preference.

c. Gender is a product of cultural and societal norms.

d. Gender is universal and fixed.

55. What are characteristics and traits considered appropriate to males and females basedon?

a. Individual preference

b. Cultural and societal norms

c. Genetics

d. Biological determinants

56. In a given society, if men are generally expected to be breadwinners and women to be caregivers, this reinforces:

a. Gender neutrality

b. Traditional gender roles

c. Equal opportunities for all

d. Gender diversity

57. In the context of gender, what does "gender identity" refer to?

a. Biological traits

b. Social roles

c. How individuals perceive themselves in terms of their gender

d. Universal gender expectations

58. Which of the following is a key factor that influences gender traits and behaviours?

a. Genetics

b. Socialization and societal expectations

c. Random chance

d. Fixed and universal standards

59. What is the term for expectations that societies hold about how individuals should behave based on their perceived gender?

a. Gender ascription
b. Gender identity
c. Gender roles
d. Gender stereotypes

60. In many societies, traditionally, which gender is often expected to be nurturing and attentive to detail?

a. Men
b. Both men and women equally
c. Neither men nor women
d. Women

61. Which definition of gender emphasizes that it is a product of the way societies categorize, differentiate, and attribute specific roles and characteristics to individuals based on their perceived gender?

a. Gender is a result of individual preference
b. Gender is a fixed and inherent quality
c. Gender is a product of biological factors
d. Gender is a social construct shaped by cultural norms

62. Which characteristic is often associated with traditional masculinity in many cultures?

a. Nurturing and caregiving
b. Attention to detail & empathy
c. Physical strength & assertiveness
d. Equality and neutrality

63. What are gender roles, behaviours, and expectations typically influenced by?

a. Biological determinants
b. Individual preference
c. Cultural and societal norms
d. Fixed and universal standards

64. What is the primary focus of the concept of "sex" in the context of biology and identity?

a. Biological and physiological characteristics
b. Societal roles and expectations
c. Personal identity
d. Cultural norms

65. What does the term "gender" primarily encompass?

a. Biological attributes
b. Social roles
c. Chromosomes
d. Hormones

66. Which term refers to the psychological and social aspects of being male or female, including gender identity, gender roles, and gender stereotypes?

a Sex
b. Gender
c. Identity
d. Chromosome

67. In the context of livestock development, why is it important to understand the conceptsof sex and gender?

a. To exclude marginalized groups
b. To combat gender-based discrimination
c. To perpetuate gender inequalities
d. To disregard cultural norms

68. How does understanding the concept of gender benefit community engagement with livestock farmers?

a. By disregarding gender norms and cultural sensitivities
b. By tailoring advice and support to gender dynamics
c. By focusing solely on individual needs
d. By promoting exclusion and discrimination

69. What does the concept of "sex" primarily refer to?

a. Socially constructed roles and behaviours
b. Biological and physiological characteristics
c. Cultural expectations
d. Psychological attributes

70. Which of the following terms is concerned with roles, behaviours, and attributes that a given society considers appropriate for women and men?

a. Gender
b. Sex
c. Culture
d. Physiology

71. In the context of gender, what is meant by "gender identity"?

a. Biological attributes that determine gender
b. The social roles assigned to individuals

c. An individual's personal perception of their gender

d. The dominant cultural norms

72. How can an understanding of gender benefit community engagement in livestock development?

a. By ignoring cultural norms and values

b. By promoting gender-based discrimination

c. By tailoring interventions to the specific needs and roles of different genders in the community

d. By excluding individuals who do not conform to traditional gender norms

73. What does the concept of "gender expression" refer to?

a. An individual's personal identity

b. Biological and physiological characteristics

c. How individuals communicate their gender to others through appearance and behaviour

d. The roles assigned by society

74. Which of the following is NOT a biological criterion for determining sex?

a. Hormones

b. Genitalia

c. Chromosomes

d. Interests

Question 2. True or False

1. Gender is solely determined by biological characteristics such as chromosomes and reproductive anatomy.
2. Gender is a social construct that encompasses societal roles, behaviours, and expectations.
3. Sex refers to psychological features and role attributions assigned by society.
4. Gender is a dynamic concept that can change over time and across different contexts.
5. Gender is a unilateral concept focusing solely on women.
6. Gender dynamics play a significant role in livestock production.
7. Gender relations within livestock production can impact the distribution of power between men and women.

8. Gender is a fixed and unchanging concept.
9. Gender is primarily determined by an individual's biological sex.
10. Gender is context- and time-specific.
11. Gender is not influenced by societal and cultural factors.
12. Gender roles and expectations can vary across different societies and cultures.
13. Gender intersects with other sociocultural factors such as class, race, and age.
14. Gender identity is typically assigned at birth based on physical characteristics.
15. Gender identity can be subject to change through gender reassignment surgeries.
16. Sex is solely about one's self-perception and identity.
17. Gender is a complex interplay of biological and cultural factors.
18. Biological sex is a binary classification.
19. Individuals with intersex variations do not conform to the binary classification of biological sex.
20. Gender is solely based on biological distinctions.
21. Gender identity can vary widely between different societies and historical periods.
22. Gender identity is fixed and unchangeable.
23. Gender expectations and stereotypes can lead to inequalities in access to resources.
24. Gender identity is a personal and individual concept.
25. Gender identity is not influenced by societal norms and expectations.
26. Gender roles can be influenced by cultural and societal norms.
27. Gender roles can shape individuals' behaviours and opportunities in society.
28. Gender roles are consistent across all societies and cultures.
29. Gender is solely about psychological characteristics.
30. Gender can be subject to change over time.
31. Gender can be shaped by cultural norms and expectations.
32. Gender is unrelated to the division of labour in society.

33. Gender roles are not influenced by the concept of masculinity and femininity.
34. Gender roles can vary between different industries and sectors, such as agriculture.
35. Gender stereotypes can influence individual behaviour and societal expectations.
36. Gender stereotypes are universal and unchanging.
37. Gender stereotypes can be challenged and transformed.
38. Gender expectations and stereotypes can lead to disparities in opportunities and access to resources.
39. Gender roles are solely based on biological distinctions.
40. Gender can intersect with other social categories like class and race.
41. Gender is a dynamic aspect of an individual's identity.
42. Gender is solely determined by one's biology.
43. Gender roles and expectations can influence decision-making authority.
44. Gender roles can be shaped by sociocultural dynamics.
45. Gender is solely about individual identity.
46. Biological sex and gender are interchangeable terms.
47. Gender can vary between different societies and cultures.
48. Gender expectations can be influenced by media and societal representations.
49. Gender expectations can lead to disparities in access to resources.
50. Gender identity is a complex interplay of societal and individual factors.
51. Gender is socially constructed and learned through socialization processes.
52. Gender roles and expectations are the same in all societies and cultures.
53. Biological characteristics that define sex are mutually exclusive and do not overlap.
54. Gender norms remain consistent across all cultures and societies.
55. Gender and sex are interchangeable terms that refer to the same concept.
56. Gender identity is solely influenced by biological factors.
57. Traditional gender norms are often hierarchical and disadvantage women.

58. Gender expression includes how an individual presents themselves through clothing and interaction with others.
59. Sex primarily refers to social and cultural aspects.
60. Biological sex can be changed through sex reassignment surgeries.
61. Gender norms are universal and consistent across all cultures.
62. Gender identity is solely determined by an individual's biological sex.
63. Gender roles are shaped by societal norms and expectations.
64. Gender stereotypes are rigid and unchanging.
65. Gender expression refers to how an individual dresses.
66. Gender norms are hierarchical and disadvantage mostly men.
67. The meaning of being male or female is fixed and universal.
68. Gender is deeply personal and varies among individuals.
69. Sex is solely determined by Society.
70. Gender is a fixed and unchanging concept.
71. Gender identity refers to how individuals perceive themselves in terms of their gender.
72. Hormones like estrogen and testosterone play a crucial role in differentiating male and female secondary sexual characteristics.
73. Gender roles and behaviours are influenced by societal expectations and cultural norms.
74. Gender diversity acknowledges and accepts a range of gender identities beyond the binary.
75. In terms of reproductive anatomy, females typically have testes.
76. Biological sex is a socially constructed concept.
77. Gender as a social construct means that gender is determined solely by biology.
78. In some cultures, pink is universally associated with boys.
79. Gender roles, behaviours, and expectations are entirely innate and biologically determined.
80. Sex is primarily shaped by individual preferences.
81. Gender is a result of socialization and societal influences.
82. Gender stereotypes are beliefs that reflect fixed and universal gender expectations
83. Sex as a social construct means that it is influenced by societal expectations and norms.

84. Chromosomes are the key determinant of an individual's gender.
85. Gender identity is the same as biological sex.
86. In terms of gender, nurture and care giving are often associated with masculinity.
87. Gender diversity is about recognizing only two distinct gender identities.
88. In many societies, males are often expected to be the primary decision-makers.
89. Traits like physical strength and assertiveness are typically associated with femininity.
90. Gender roles are shaped by cultural and societal norms.
91. Gender roles and expectations are universal and unchanging across cultures.
92. Gender is entirely determined by individual choice.
93. Biological factors are the sole determinants of gender.
94. Traditional gender roles can vary widely between different societies.
95. Biological sex and gender identity are always aligned.
96. Gender is a concept that is free from societal influences.
97. Gender diversity is important in promoting inclusivity and respect for all gender identities.
98. Gender identity is a fixed and unchanging attribute.
99. Gender is only about individual identity, not social groups.
100. In some cultures, blue is universally associated with girls.
101. Gender roles are entirely shaped by biology.
102. Gender diversity is the recognition of a spectrum of gender identities.
103. In many societies, males are often expected to excel in nurturing and caregiving roles.
104. Sex and gender are interchangeable terms that mean the same thing.
105. Sex is determined by biological characteristics such as chromosomes, reproductive organs, and hormones.
106. Gender-based discrimination is a pervasive issue that can hinder livestock-based economic development.
107. Gender inclusivity in livestock development initiatives is important to address the needs and rights of all individuals.
108. Biological sex and gender are always aligned, with no exceptions.

109. Gender identity is a social construct.
110. Gender is a static and unchanging aspect of an individual's identity.
111. In most societies, there are no differences or inequalities between women and men in terms of responsibilities and access to resources.
112. Sex is a binary concept, limited to notions of masculinity and femininity.
113. Sex reassignment surgeries and hormone therapy can change an individual's gender.
114. Understanding the distinction between sex and gender is irrelevant in the field of livestock development.
115. Gender is solely determined by biology, with no cultural or societal influence.
116. Sex and gender are both complex and multifaceted concepts.
117. Gender identity is typically recognized and understood in the same way by all individuals.
118. Gender norms and expectations are the same in all societies and cultures. Understanding the distinction between sex and gender is irrelevant in the field of community engagement.
119. Gender identity is primarily determined by an individual's biological characteristics.
120. Gender is a fixed and unchanging aspect of an individual's identity.
121. Gender and sex are interchangeable terms that refer to the same concept.
122. Gender identity is solely influenced by biological factors.
123. Gender norms remain consistent across all cultures and societies.
124. Gender expression is a personal and internal aspect of an individual's identity.
125. Gender refers only to the biological differences between males and females.
126. Cultural sensitivity is not necessary when working with communities on livestock development.
127. In some cases, individuals may identify as non-binary or genderqueer, reflecting a wide range of gender identities.
128. Gender intersects with other social categories such as class, race, and ethnicity.
129. Gender expression is how individuals communicate their gender to others through clothing and body language.

130. Gender inclusivity can help promote economic empowerment in livestock-based economic development.
131. In some societies, women may face challenges in accessing loans or modern farming equipment for livestock development.
132. Gender roles are culturally constructed and can vary across different societies.
133. Hormones play a crucial role in the differentiation of male and female reproductive structures.
134. Traditional gender norms are often hierarchical and disadvantage women.

Question 3. Fill in the Blanks

1. Gender refers to the socially constructed differences associated with being a _______ or a _______.
2. Gender is the outcome of a social process through which differences based on _______ are defined, imagined, and become significant in specific contexts.
3. Gender is changeable over time and across _______.
4. Gender is a dynamic relationship between _______ and _______.
5. Access to and control over livestock is one of the key areas of consideration when exploring gender dimensions in livestock-based systems, and it represents gendered _______.
6. Gender imbalances contribute to undermining and/or promoting livestock development efforts and can have consequences, especially in _______ contexts.
7. Sex encompasses the biological characteristics that typically categorize individuals as _______, _______, or _______ based on physical attributes.
8. Chromosomes, hormones, and reproductive anatomy are components of _______ that differentiate males and females.
9. Females typically have _______ chromosomes, whereas males possess _______ chromosomes.
10. Key hormones associated with each gender are _______ in females and _______ in males.
11. Gender is not necessarily restricted to the binary classification of male or female but can encompass a range of _______ and _______.

12. Gender roles and expectations are influenced by societal norms, and they may shape roles within social groups, even in the absence of significant _______ differences.
13. Gender _______ play a crucial role in shaping the division of labour on farms and other aspects of livestock production.
14. In many societies, certain tasks are associated with traditional gender expectations, such as tasks like milking cows being labelled as _______ and caring for animals.
15. Gender roles can vary between different _______ and change over time.
16. Gender intersects with other social categories, including _______, _______, and _______.
17. Gender is a complex and dynamic aspect of an individual's _______, continually evolving and shaping their experiences within society.
18. Gender is learned through the process of _______ and varies across different _______.
19. Sex refers to the biological and physiological characteristics that define humans as _______, _______, or _______.
20. Gender refers to socially constructed roles, behaviours, activities, and attributes that society considers appropriate for _______ and _______.
21. Gender is a complex and multifaceted concept that encompasses social, psychological, and cultural _______.
22. Gender influences what is considered appropriate and valued for individuals of a particular _______.
23. Gender is not determined by _______, but is constructed and learned through social and cultural processes.
24. In the context of livestock production, gender relations are highly important and are concerned with the systematic differentiation of men and women in processes of _______.
25. Gender _______ are pervasive in different ways across many aspects of livestock development.
26. Gender influences the distribution of power between the sexes in livestock production, as communities often use gender as a category of social differentiation to delineate differences between men and women and to justify, sustain, and reinforce _______.
27. In the context of livestock production, gender roles and expectations may shape the division of labour on farms, even in the absence of significant _______ differences.

28. Gender roles and expectations can influence individual behaviour and societal expectations, including clothing choices, interests, and roles within family and community _______.
29. Gender is not a unilateral concept focusing solely on women but is the dynamic relationship between _______ and _______.
30. Understanding the nature of gender dynamics is significant if livestock-based interventions are to sustainably improve the overall welfare of men and women and limit any unintended _______ for them.
31. In the context of livestock-based interventions, gendered access to and control over _______ is an important consideration.
32. Gendered divisions of labour in livestock management may result from societal expectations related to traditional _______ roles.
33. Gendered access to livestock technologies may be influenced by _______ norms and expectations.
34. Access to livestock extension and training can be influenced by perceived _______ roles and capabilities.
35. Gendered intra-household dynamics regarding the distribution of costs and benefits from livestock and livestock interventions can reflect societal norms related to the _______ of responsibilities.
36. Gender is a socially constructed concept that goes beyond the confines of biological _______.
37. Gender is depicted as a product of societal expectations, norms, and _______ influences.
38. In agriculture, tasks like milking cows may be associated with nurturing and caregiving, typically labelled as _______ traits.
39. Handling equipment maintenance might be linked to physical strength, traditionally seen as _______.
40. Gender is part of the broader ______ context and is influenced by factors such as class, race, and age.
41. Sex is primarily defined by ________, ________, and ________ aspects.
42. Gender is a social, psychological, and cultural construct that develops during the process of ________.
43. Gender norms vary across different ________ and over time.
44. Gender intersects with other categories such as ________, ________, and ________.

45. Gender expectations are learned through socialization processes within ________, ________, and ________.
46. The European Institute for Gender Equality defines gender as the social attributes and opportunities associated with being ________ and ________.
47. Gender expression refers to how an individual ________ their gender.
44. ________ is a social construct that varies across different societies.
45. Gender roles and expectations are primarily shaped by ________ norms and ________.
46. Gender norms are often limited to notions of ________ and ________.
47. Sex is primarily determined by ________________, such as chromosomes, reproductive organs, and hormones.
48. Gender refers to socially constructed roles, behaviors, and attributes that a given society considers appropriate for ________________ and ________________.
49. Gender identity is an individual's personal perception of their own ________________.
50. Gender norms can vary across different __________ and __________.
51. Sex and gender are not always aligned, and there are individuals who may identify as ________________ or ________________.
52. Gender expression refers to how individuals communicate their gender to others through ________________ and ________________.
53. In most societies, there are differences and inequalities between ________________ and ________________ in responsibilities, access to resources, and decision-making opportunities.
54. Gender intersects with other social categories such as ________________, ________________, and ________________.
55. Understanding the distinction between sex and gender is essential for promoting ____________ and ____________ in development initiatives.
56. Gender-based discrimination is a pervasive issue that can hinder economic ________________.
57. Gender inclusivity in community engagement is crucial to address the needs and rights of all ________________.
58. Cultural sensitivity is necessary when working with communities to understand and respect their ________________.

59. Gender identity is not necessarily determined by ______________.
60. Gender roles and expectations are socially constructed and can vary over ______________.
61. Sex reassignment surgeries and hormone therapy can change an individual's ______________ but not their ______________.
62. Gender norms often revolve around notions of ______________ and ______________.
63. Gender expression can include choices related to ______________, ______________, and ______________.
64. Gender is both an analytical category and a political idea that addresses the distribution of power in ______________.
65. Gender identity can be deeply personal and may be recognized early in childhood or ______________.
66. Gender is not limited to a binary notion and can encompass a wide range of ______________.
67. Understanding sex and gender helps create strategies that promote economic ______________ for all.
68. Gender norms and expectations can influence roles and responsibilities within the ______________.
69. Gender is developed through ______________ and is context- and time-specific.
70. Inclusive development initiatives consider the rights and needs of all individuals, irrespective of their ______________ or ______________.
71. Gender identity is deeply personal and can be recognized at different stages of ______________.
72. Some societies create norms and expectations based on a heteronormative order that assumes ______________ genders.
73. Gender identity may or may not correspond to an individual's ______________.
74. Gender inclusivity is important to combat gender-based ______________.
75. Understanding gender is relevant when working with livestock farmers to address issues related to ______________ and ______________.
76. Gender norms can change over time but are often based on a ______________ framework.

77. In the differentiation of males and females, sex encompasses the interplay of ______, ______, and ______.
78. Gender "refers to the social categories of ______ and ______."
79. The American Psychological Association (APA. suggests that gender is cultural and is the term to use when referring to women and men as ______ groups.
80. Gender, as a social construct, goes beyond the ______ category of sex.
81. Gender is not solely determined by physical characteristics but is also shaped by ______ expectations and norms.
82. Gender is not a fixed or inherent quality but a socially constructed ______.
83. The APA's definition of gender acknowledges that gender is deeply rooted in ______ norms, roles, and expectations.
84. According to Helgeson, gender is distinguished by a set of ______ features and role attributions.
85. The ______ characteristics assigned by society play a role in defining gender.
86. The American Psychological Association (APA)emphasizes that gender is a term used when referring to women and men as ______ groups.
87. Gender is not solely an individual characteristic but something deeply rooted in ______ and societal norms.
88. Society assigns specific ______ features and roles based on biological sex.
89. The differences between men and women are shaped by ______ expectations and norms.
90. Gender, according to the APA, is deeply rooted in cultural, historical, and ______ factors.
91. Gender is a concept that is influenced by cultural, historical, and ______ factors.
92. Gender, as a social construct, is not solely determined by physical ______.
93. Gender is a term used when referring to women and men as ______ groups, according to the APA.
94. Society assigns specific ______ based on an individual's biological sex.
95. Gender goes beyond the biological category of ______.

96. Gender is influenced by ______ norms, roles, and expectations.
97. The American Psychological Association (APA)acknowledges that gender is not solely an individual characteristic but something that is deeply rooted in cultural, historical, and ______ factors.
98. Gender, according to Helgeson, is distinguished by a set of ______ features and role attributions assigned by society.
99. The __________ characteristics assigned by society play a role in defining gender.
100. The differences between men and women are shaped by ______ expectations and norms.

ANSWERS

1. Multiple Choice Questions

1	b	Social roles, responsibilities and expectations based on perceived gender identity
2	a	Sex is about biological characteristics; Gender is a social construct
3	c	Milking cows
4	b	Economic, social, and technological interactions in animal farming
5	b	They can lead to unintended consequences and inequalities.
6	b	Access to and control over livestock
7	c	Men handling heavy machinery maintenance
8	d	Gender stereotypes
9	b	It's a binary classification based on chromosomes and hormones.
10	d	Gender is about social roles and behaviors, while sex is about biological attributes.
11	b	Social categories distinguished by psychological features
12	a	Through clothing choices, interests, and roles
13	c	Social and cultural norms
14	b	Psychological and social aspects of being male or female
15	b	It represents social and cultural influences on males and females.
16	d	Gender intersects with other social categories like class and race.
17	b	The result of learned behaviors and cultural norms
18	b	Social roles, responsibilities, and expectations based on perceived gender identity

19	c	Economic and social interactions
20	b	It reinforces and sustains power imbalances.
21	b	Social roles, responsibilities, and expectations based on perceived gender identity
22	c	Hindering the development of livestock-based livelihoods
23	c	Power
24	b	Sex
25	c	Gender encompasses social attributes and roles associated with being male or female.
26	a	Gender refers to biological differences, while sex is a social construct.
27	c	Gender is shaped by cultural norms and can change over time.
28	d	Gender refers to psychological traits of masculinity and femininity.
29	c	How an individual feels about themselves in terms of masculinity or femininity
30	d	Non-binary or genderqueer
31	a	Biological and physiological aspects
32	a	Chromosomes, hormones, and reproductive anatomy
33	c	How an individual communicates their gender
34	b	Socialization and cultural influences
35	c	Gender stereotypes
36	c	Societal norms and culture
37	d	Gender roles
38	b	Heteronormativity
39	b	Chromosomes
40	c	Shaped by societal expectations and norms
41	c	Boys are better at math, and girls are better at language.
42	b	The influence of society and culture in shaping gender roles and behaviors
43	d	Gender ascription
44	d	Socially constructed
45	c	Based on societal expectations
46	c	Chromosomes
47	c	Testosterone
48	b	Uterus and ovaries

49	d	Gender identity
50	b	Masculinity
51	c	Cultural and societal norms
52	b	It signifies that gender is shaped by societal expectations and norms
53	b	Physical strength and assertiveness
54	c	Gender is a product of cultural and societal norms.
55	b	Cultural and societal norms
56	b	Traditional gender roles
57	c	How individuals perceive themselves in terms of their gender
58	b	Socialization and societal expectations
59	c	Gender roles
60	d	Women
61	d	Gender is a social construct shaped by cultural norms
62	c	Physical strength and assertiveness
63	c	Cultural and societal norms
64	a	Biological and physiological characteristics
65	b	Social roles
66	b	Gender
67	b	To combat gender-based discrimination
68	b	By tailoring advice and support to gender dynamics
69	b	Biological and physiological characteristics
70	a	Gender
71	c	An individual's personal perception of their gender
72	c	By tailoring interventions to the specific needs and roles of different genders in the community
73	c	How individuals communicate their gender to others through appearance and behavior
74	d	Interests

Answer 2. True and False

S.No.	True or false	S.No.	True or false
1	False	2	True
3	False	4	True
5	False	6	True

7	True	8	False
9	False	10	True
11	False	12	True
13	True	14	False
15	False	16	False
17	True	18	True
19	True	20	False
21	True	22	False
23	True	24	True
25	False	26	True
27	True	28	False
29	False	30	True
31	True	32	False
33	False	34	True
35	True	36	False
37	True	38	True
39	False	40	True
41	True	42	False
43	True	44	True
45	False	46	False
47	True	48	True
49	True	50	True
51	True	52	False
53	False	54	False
55	False	56	False
57	True	58	True
59	False	60	True
61	False	62	False
63	True	64	False
65	True	66	False
67	False	68	True
69	False	70	False
71	True	72	True
73	True	74	True

75	False	76	False
77	False	78	False
79	False	80	False
81	True	82	True
83	False	84	False
85	False	86	False
87	False	88	True
89	True	90	True
91	False	92	False
93	False	94	True
95	False	96	False
97	True	98	False
99	False	100	False
101	False	102	True
103	False	104	False
105	True	106	True
107	True	108	False
109	True	110	False
111	False	112	False
113	False	114	False
115	False	116	False
117	False	118	False
119	False	120	False
121	False	122	False
123	False	124	False
125	False	126	False
127	True	128	True
129	True	130	True
131	True	132	True
133	True	134	True

Answer 3. Fill in the Blanks

S.No.	Answers
1	Man, woman
2	Biological sex
3	Contexts
4	Men, women
5	Imbalances
6	Rural
7	Male, female, intersex
8	Sex
9	XX, XY
10	Estrogen, testosterone
11	Identities, expressions
12	Biological
13	Norms
14	Feminine
15	Societies
16	Class, race, age
17	Identity
18	Socialization, cultures
19	Female, male, intersex
20	Women, men
21	Constructs
22	Sex
23	Biology
24	Production
25	Imbalances
26	Inequalities
27	Biological
28	Dynamics
29	Men, women
30	Consequences
31	Productive resources
32	Gender

33	Cultural
34	Gender
35	Division
36	Sex
37	Cultural
38	Feminine
39	Masculine
40	Social
41	Biological, physiological, anatomical
42	Socialization
43	Cultures
44	Class, skin color, ethnicity
45	Family, school, media
46	Female, male
47	Biology
48	Men, Women
49	Societal, culture
50	Masculinity, femininity
51	Intersex; non-binary
52	Appearance; behavior
53	Women; men
54	Class; race; ethnicity
55	Inclusivity; equality
56	Development
57	Individuals
58	Cultural norms
59	Biological sex
60	Time
61	Sex; gender identity
62	Masculinity; femininity
63	Clothing; hairstyle; body language
64	Society
65	Later on
66	Identities

67	Empowerment
68	Community
69	Socialization processes
70	Gender; sex
71	Life
72	Two
73	Biological sex
74	Discrimination
75	Access to resources; decision-making
76	Heteronormative
77	Chromosomes; hormones; reproductive anatomy
78	Male; female
79	Social
80	Biological
81	Societal
82	Concept/Construct
83	Cultural
84	Psychological
85	Role
86	Social
87	Cultural
88	Psychological
89	Societal
90	Societal
91	Societal
92	Characteristics
93	Social
94	Roles
95	Sex
96	Societal
97	Societal
98	Psychological
99	Role
100	Societal

2

Women's Role in Livestock Production Empowerment and Challenges

Chapter Overview

In this chapter, we delve into the pivotal role of women in livestock production, particularly in the context of India. Women emerge as the backbone of livestock care, contributing significantly to rural communities' well-being and the nation's economy. We explore their diverse and indispensable contributions, from feeding and general management to health and breed management, processing, and marketing. Despite their crucial roles, women face numerous obstacles and constraints in livestock production, including limited access to resources, gender biases, and cultural norms.

The chapter also highlights the underestimated scale of women's involvement in livestock production, emphasizing the need to recognize and support their contributions fully. We discuss the empowerment of women through livestock ownership, including its benefits for decision-making, food security, income generation, and self-esteem. Additionally, we outline measures to promote women's empowerment in livestock production, addressing obstacles, and offering strategies for change.

Overall, this chapter sheds light on the unsung heroines of the fields, the women who nurture and care for livestock, and the importance of empowering them to enhance the sustainability and success of rural communities and the well-being of the animals they tend to.

SECTION A: THEORY

Role of Women in Livestock Production

Women's role in livestock sector in India is of paramount importance, given the country's heavy reliance on agriculture and the livestock industry. In a nation where approximately 8.80% of the population is employed in the livestock sector, the significance of this sector cannot be overstated. It serves as a lifeline for two-thirds of the rural population, contributing 4.11% to the overall GDP and a substantial 25.60% to the agricultural GDP. Livestock not only provides a livelihood for around 20.50 million people but also constitutes 16% of the income of small farm households and benefits 14% of total rural households. (Livestock Census 2012).

The cow produces dung, milk and offspring, much of which is managed in a way or another by the women of the family

Source: Chakravarty, R. and J.P. Dhaka. (1995). Delving into gender analysis. In: Handbook for Straw Feeding Systems (Edited by Kiran Singh and J.B. Schiere), pp. 20-25. ICAR, New Delhi, India.

Within the context of India's animal husbandry sector, women emerge as the primary workforce, and their contributions are diverse and indispensable. Several key statistics underscore the pivotal role of women in agriculture and livestock production:

i. Livestock are critically important for women and women are critically important for livestock
ii. Globally, it is estimated that about two-thirds of impoverished livestock keepers, totalling approximately 400 million people (out of 600 million), are women
iii. The agriculture sector boasts the highest percentage distribution of female workers, closely followed by manufacturing in India.
iv. Nationwide, nearly 63% of workers in the agriculture sector are female.
v. Women comprise an estimated 71% of the labour force in livestock farming.
vi. In India, according to the Employment and Unemployment Survey (EUS) 2012, around 8 million female workers and 2 million male workers are engaged in livestock rearing in the rural agricultural workforce.

vii. In India, a staggering 75 million women are engaged in dairying, compared to 15 million men.

viii. The number of women members in dairy co-operatives has increased from 5 million in 2015-16 to 5.4 million in 2020-21.

ix. Women now account for 31% of all members of dairy producer cooperatives, with the number of women's dairy cooperative societies rising from 18,954 in 2012 to 32,092 in 2015-16.

x. This expansion is also evident in the proliferation of women's dairy cooperative societies in India, which surged from 18,954 in 2012 to an impressive 32,092 in 2015-16.

xi. In states like Punjab and Haryana, where dairying is prominent and animals are stall-fed, the share of women employment in the livestock sector is as high as 90%.

xii. On average, women dedicate 3 to 5 hours a day to livestock production activities.

xiii. Female farmers invest a significant amount of time in dairy activities, dedicating approximately 1,779 hours per year, compared to 315 hours per year for men.

xiv. Unfortunately, women often bear the physical toll of their engagement in livestock activities, with 55.3% of female farmers experiencing musculoskeletal disorders (MSD), including back pain.

xv. 95% of female livestock workers are self-employed. There is a minimum presence of casual workers in livestock raising.

xvi. Rural women engage in multiple activities depending on seasons and availability of work. Livestock raising is one among them usually carried out in the homestead by the household members.

xvii. Regional variations in women's participation in livestock raising exists with higher participation rates in the Northern States of India, as well as in Gujarat, and Kerala.

xviii. According to a report by the Food and Agriculture Organization (FAO 2011) if women were to have access to the same level of resources as men, agricultural productivity would go up by 10–30 per cent and agricultural output would increase by up to 4 per cent".

xix. Women who access and control livestock assets improve the health, education and food security of their households. 90% of income under the control of women is channelled back into their households or local communities, compared with only 30-40% for men.

xx. Inclusive livestock development can greatly advance the achievement of SDG5 [UnitedNations Sustainable Development Goals Number

5(Achieve gender equality and empower all women and girls] and foster gender equality and the empowerment of women, particularly in rural areas.

Women in Livestock production

The role of women in different livestock production activities is crucial and multifaceted. Women are actively involved in various aspects of livestock management, contributing significantly to rural communities and ensuring the well-being of the animals they care for.

1. **Feeding Management**: Women in hilly regions often collect and transport fodder from distant areas. They possess valuable knowledge of local feed resources, including grasses, weeds, and feed trees, essential for animal nutrition. Young girls also participate in grazing small ruminants. Women spend a substantial amount of time, approximately 132.07 hours per year, cutting grass and collecting food for the animals ((Kishtwaria *et al.*2009). In regions like Karnataka and Gujarat, high percentages of women are engaged in activities such as feeding, providing water, and storing food for animals, underscoring their critical roles in livestock care. Some of the feeding-related activities undertaken by women include feeding cattle, collecting fodder, taking animals for grazing, chaffing fodder, preparing feed, and creating homemade concentrate mixtures.
2. **General Management**: Women dedicate significant time, approximately 294.34 minutes each day (Thirunavukkarasu & Christy, 2002) to various daily dairy activities which, include watering, milking, cleaning livestock and their stalls, and preparing dung cakes.
3. **Health and Breed Management**: Women play essential roles in caring for sick animals, managing pregnant animals, looking after milch cattle, providing traditional healthcare, caring for new-born calves, and attending to animals during parturition.
4. **Processing and Marketing**: Women are primarily involved in processing activities related to dairy production, such as selling milk, creating value-added dairy products (curd, butter, milk sweets etc), storing dairy products, and marketing these products. They also engage in trading livestock especially small ruminants.

   However, there are certain areas within livestock production where women have limited roles, including decision making, taking loans for livestock production, purchasing and selling animals, choosing animals for dairy purposes, and having less contact with progressive farmers, officials, and banks. For instance, in Rajasthan, women work approximately 9.4 hours a day, with their time divided among household chores, animal care, and spinning (Social Assessment Report, 2010). While women are actively engaged in small-scale processing and selling dairy products, their participation in financial endeavours, such as maintaining business records and securing bank loans, receives less attention.

Overall, recognizing and supporting the diverse roles that women play in livestock production is essential for enhancing the sustainability and success of rural communities and ensuring the well-being of the animals they care for. Efforts to empower women in various aspects of livestock management can lead to increased productivity and improved livelihoods in these regions.

Additional Notes

1. The contribution of women in livestock related activities is well recognised.Now the question is that they are not treated on par with men on various aspects which include:
   i. Access to land and livestock resources
   ii. Access to information on livestock production
   iii. Access to training to improve their capacities
   iv. Access to marketing of livestock and livestock products
   vi. Participation in decision making on various livestock development related issues like sale / purchase of animals, what technology to be adopted, when to approach and whom to approach when they are experiencing a problem in livestock related aspects,
   vi. Opportunity to express their opinions, for example in matters of improving the production of livestock etc.
2. When the women have equal opportunities like men in improving their capacity in attending to different activities related to both household and livelihood activities, their contribution to the overall welfare of the families will go up which will further empower them to enhance their contribution at society as well as national level.

The Unsung Heroines of the Fields: Women Livestock Farmers Nurturing Lives

Meet Radha, a spirited woman with a deep connection to the land and its creatures. Each day, she ventures out into the emerald landscapes, collecting fodder for her cattle. With a keen eye, she identifies valuable grasses, weeds, and feed trees, knowledge passed down through generations. Radha knows that the health and well-being of her animals depend on the nourishment she provides, and she takes this responsibility seriously.

As she walks the rolling hills, Radha's young daughter, Meera, trails behind, learning the ropes from her mother. Meera is just one of many young girls in the region who participate in the age-old tradition of grazing small ruminants. They absorb wisdom about the land and its bounty, knowledge that will serve them and their communities well in the years to come.

In their home state of Karnataka, Radha and her fellow women farmers are the backbone of livestock care. As per research findings (Rathod *et al.*, 2011), about 87% of women, like Radha, tend to their animals' needs. They ensure their cattle have clean water to drink, help them find suitable grazing spots, and meticulously cut grass and collect food every two weeks—a task that consumes about 132 hours each year.

Travel a bit west to the state of Gujarat, and you'll find women like Priya, who are equally devoted to their livestock. Approximately 81% of female farmers in the region engage in collecting food for their animals, while 75% take on the responsibility of chopping the fodder. Priya is among the 87% who personally deliver the nourishment to their animals, ensuring they receive the care they need. With dedication, they store food and water, safeguarding their animals against the harshness of nature.

Beyond feeding, these women excel in a myriad of tasks essential for livestock care. They dedicate nearly five hours each day to activities such as watering, milking, cleaning stalls, bathing and grooming animals, housing them, and even gathering cow dung for practical use. Their tireless efforts ensure that their livestock is healthy, productive, and comfortable.But it doesn't stop there. In the intricate dance of livestock management, women are also the caregivers, tending to sick animals, nurturing pregnant ones, and supporting milch cattle. Their traditional knowledge about healthcare and calving is invaluable, passed down through generations.

As the day unfolds, you'll find women like Radha and Priya not only caring for their animals but also engaging in processing activities, such as selling milk, preparing dairy products, and storing these precious resources. Their homes become centres of dairy production, where they transform raw milk into curd, ghee, and butter milk, adding value to their livelihoods.

However, not all aspects of livestock management fall under their purview. When it comes to financial endeavours, such as receiving sale proceeds of milk or animals securing loans and managing business records, women's participation is often limited. Yet, their resilience and determination are unwavering.

These women, in the backdrop of rolling hills and verdant pastures, form the backbone of their rural communities. Their tireless efforts ensure the well-being of their animals and the sustenance of their families. They are the embodiment of strength, wisdom, and dedication, and their contributions to Indian agriculture deserve recognition and support. In the heartland of India, these women farmers stand as a testament to the power and resilience of those who till the land and nurture the animals that sustain us all.

Empowering Women through Livestock Ownership

Livestock ownership offers a range of significant benefits to women, including enhanced decision-making power, improved household welfare, increased income generation, and heightened self-esteem. Here's a more streamlined breakdown of these advantages:

1. **Empowerment and Decision-Making:** Women who own livestock gain influence within their households and communities. They actively participate in critical decisions related to livestock, such as sales, disease management, and animal selection, ultimately leading to more equitable decision-making processes.
2. **Food Security:** Livestock ownership to women ensures a consistent supply of nutritious animal-based protein-rich products like milk, meat, and eggs, benefiting the entire family and addressing food security concerns.
3. **Income Generation:** Livestock especially dairy cattle serves as a reliable and regular source of cash income for women. The proceeds from livestock sales can be used for various purposes, such as medical expenses and education fees, ultimately contributing to poverty reduction.
4. **Enhanced Household Welfare:** Livestock management ultimately generate substantial income, contributing significantly to the overall well-being of the household.
5. **Self-Esteem:** Active involvement in livestock-related activities bolsters women's self-esteem and recognition as valuable contributors to their households and communities. Women having livestock are more respected than those who do not have.

6. **Asset Ownership:** Livestock, compared to land or financial assets, are more accessible for women. However, safeguarding women's ownership rights is crucial to empower them further, enhancing their bargaining power and decision-making roles within the household. Livestock when in financial distress serve as a guarantee to take loans from money lenders or milk vendors or middlemen in animal trade in the villages
7. **Income Management:** Effective income management by women is essential for the survival of many households. When women control income, it often leads to improved resource allocation, reduced instances of domestic violence, and better child nutrition. Notably, women can manage income from livestock products, even when they don't own the livestock themselves, underscoring their vital role in income management. Women spend money from livestock rearing on meeting the family requirements (food, clothing, school fee, health etc) contrary to men who spend money on other purposes.

   Empowering women through livestock ownership not only benefits them individually but also has a cascading impact on their families and communities, fostering inclusive and sustainable development. It's imperative to support initiatives that promote women's access to livestock resources, enabling them to harness these advantages fully.

Constraints faced by women in livestock production

In many developing countries, women engaged in agriculture and animal husbandry encounter numerous challenges, negatively impacting agricultural productivity. The Food and Agriculture Organization (FAO) noted in 2011 that agriculture is not performing optimally in these regions, partly due to women lacking the necessary resources and opportunities to efficiently utilize their time. Despite being farmers, workers, and entrepreneurs, women consistently face more significant constraints than men when it comes to accessing productive resources, markets, and services. This gender disparity, often referred to as the "gender gap," hampers the productivity of women in the agriculture sector and diminishes their overall contributions to broader economic and social development objectives.

According to the FAO report, addressing gender inequalities in access to productive resources and services has the potential to yield significant benefits. Specifically, reducing these disparities could result in a notable increase in yields on women's farms, ranging from 20 to 30 percent. This, in turn, could lead to a substantial enhancement of agricultural output in developing countries, with estimates suggesting an overall increase of 2.5 to 4 percent. Women involved in livestock production face numerous constraints that hinder their empowerment and active participation in this sector. These constraints include.

1. **Insecure Land Tenure**: Women often struggle to access and own land, directly affecting their engagement in livestock production. It is essential to ensure secure land tenure for women to empower them in this sector.
2. **Limited Access to Resources:** Women typically have restricted access to credit, extension services, and modern technologies, significantly impeding their productivity and income potential. Women frequently face barriers in accessing productive resources like land, livestock, credit, and technology, limiting their full participation and benefits from the livestock sector. Providing women access to these resources is crucial for empowerment. In some communities' women have no access on household cash income
3. **Limited Control over Assets**: Women may lack control over the income generated from livestock production or ownership of the animals themselves. Empowering women with ownership and control over livestock assets can enhance their decision-making power and economic status.
4. **Gender Roles and Labour Division**: Gender-specific roles and responsibilities are often conditioned by household structures, accessible resources, specific impacts of the global economy and other locally relevant factors, such as ecological conditions (FAO, 1997). Traditional gender roles often confine women to specific tasks within livestock production, such as cleaning, feeding, milking, processing of livestock products, and marketing. This results in heavy workloads for women and constrains opportunities for diversification and livelihood improvement. Efforts should be made to reassess and adjust the division of labour to reduce the risk of drudgery and create more opportunities for women.
5. **Intensification of Livestock Production**: The growing urban demand for livestock products is increasingly met by intensive commercial systems, which may further marginalize women's participation in the sector. In the recent past large scale commercial dairy farms are on the rise where the scope of women involvement is less.
6. **Limited Access to Extension, Marketing, and Credit Services**: Men often have greater mobility and responsibility for activities like breeding and fodder production, leading to their more frequent interactions with extension agents, animal health specialists, and input suppliers. The active role that women play in livestock care is often overlooked in these services, exacerbating gender disparities.

7. **Gender Biases and Cultural Norms**: Societal norms and gender biases frequently restrict women's mobility and limit their involvement in livestock-related activities, hindering their access to training, information, and market opportunities. Cultural attitudes that undervalue women and their contributions pose a significant constraint. However, change is possible through economic and social incentives, the presence of role models, policy reforms, and exposure to alternative options.
8. **Lack of Decision-Making Power**: Women often have limited decision-making power in livestock-related activities, including breeding, feeding, and marketing. Their contributions are often undervalued and not recognized in national policies and plans.
9. **Lack of Institutional Support**: The livestock sector requires more robust institutional support to address gender disparities, including providing credit facilities, training programs, and upgrading women's livestock farming skills.
10. **Access to capital**: Men usually find it easier to get loans from the government because they can offer something valuable as a guarantee. Women often can't read or write, and they may need their husband's permission to deal with officials.
11. **Veterinary Services:** In many countries in Asia, Africa, and Latin America, services for taking care of animals are mostly focused on helping men. Vets and experts usually work with men and their animals. Training of men in the areas where women are involved serves limited purpose in improving the livestock production. It is also observed that men do not pass the information received during training to their counterparts.
12. **Extension Services**: Extension personnel who teach about animals aren't always good at explaining things to women. They might not understand the language women speak, and women might not be able to read. Often, they first help men because they think it's more important. For extension education for women special skills may be needed as teaching women about animals is a bit different and difficult for men extension specialists. You need special skills because women often speak local languages, can't read, and need simpler explanations. Unfortunately, in many countries' women extension specialists are very limited and it is impossible to cater to the diverse requirements of the women livestock farmers. It was estimated that the women extension agents in the world were only 15 percent and male extension agents frequently focus on male-farmers (World Bank/FAO/IFAD, 2009).

The Underestimated Role of Women in livestock production

The extent of women's contributions to animal care on farms often goes unrecognized. Official surveys suggest that approximately 12 million rural women are engaged in livestock raising. However, a closer examination by researchers reveals a starkly different reality – around 49 million rural women are actively involved in this critical work. This means that the actual number of women participating in livestock production is four times greater than what official surveys indicate, constituting a significant portion of the rural population. Recent surveys in India corroborate this trend, revealing that roughly 11% of rural women, equivalent to about 48 million individuals, play a role in caring for animals in various capacities.

13. **Unpaid Family Labour:** Many women take care of animals without getting paid. This means they do it as part of their family duties. The money earned from selling animal products goes to household income, but it may not always go directly to the woman doing the work.

    In conclusion, addressing these obstacles and constraints is imperative for promoting gender equity in livestock production and harnessing the full potential of women in this sector. Empowering women in livestock production not only benefits them individually but also contributes to overall agricultural and economic development.

Decline in women's participation in livestock raising in India since 2004-05

The possible factors that have contributed to the decline in women's participation in livestock raising since 2004-05 include:

i. **Changing preferences**: younger women, especially those aged 30 and below with education, are less interested in traditional activities like livestock raising. They prefer to engage in other income-generating activities, such as tailoring.

ii. **Decline in cattle and buffalo ownership**: There has been a decrease in the number of cattle and buffaloes owned by rural households since 2002-03. This decline in ownership may have led to a decrease in female participation in raising animals.

iii. **Regional variation**: Women in Northern States participate more in raising bovines, and their participation in livestock raising has been above the all-India average.

Measures to promote women empowerment in livestock production

Empowering women in rural areas to have a greater role in decision-making activities related to livestock production requires addressing several key factors. Here are some strategies that can be implemented:

1. **Ensure Secure Land Tenure for Women**: Implement policies and programs that secure and expand women's access to and control over land, tailoring strategies to the region and society's specific context.
2. **Enhance Access to Resources**: Ensure women have equal access to productive resources like land, livestock, credit, and technology through targeted policies and initiatives promoting ownership and control. Provide women with access to credit, extension services, training, and technologies. Specialized credit lines for women can be successful if they are transparent and culturally adapted. Ensure women have access to relevant information, technologies, and resources critical for empowerment. This can be achieved through extension services, advisory support, and access to credit and financial services, enabling informed decision-making.
3. **Promote Ownership and Control of Livestock Assets**: Develop strategies to empower women to own and control livestock, analysing project area conditions and setting realistic goals while monitoring progress.
4. **Address Gender Roles and Labour Division**: Analyse the roles and responsibilities of men and women in livestock production and adjust the division of labour accordingly. Introduce labour-saving measures to alleviate women's heavy workloads and create diversification opportunities.
5. **Capacity-building Programs**: Implement capacity-building programs to equip women with essential knowledge and skills in livestock management, including specific animal husbandry practices and technical expertise. Offer targeted training and skill development programs for rural women, covering aspects like animal health, nutrition, breeding, and management practices. Equipping women with these skills enables them to actively engage in decision-making.
6. **Strengthen Women's Networks and Organizations**: Encourage formation of women's clubs, cooperatives, self-help groups, and organizations to create platforms for women to share experiences and collectively address challenges. These networks offer opportunities for training, capacity-building, and market access, enhancing women's participation in decision-making.

7. **Promote Gender Equality and Challenge Social Norms**: Tackle gender biases and promote equality within households and communities through campaigns, sensitization programs, and advocacy to ensure equal rights and opportunities for women in livestock production. Challenge and change gender biases and cultural norms that hinder women's participation in the livestock sector. Employ awareness campaigns, community dialogues, and targeted interventions to promote gender equality and women's empowerment.
8. **Engage Men in the Process**: Involve men in discussions and negotiations related to livestock production to create a supportive environment for women's empowerment. Partner with men as allies in decision-making, ensuring women's voices are heard and respected.
9. **Promote Women's Participation in Decision-Making**: Encourage and support women to actively engage in decision-making processes related to livestock management. Create platforms for their voices, offer leadership and negotiation skills training, and promote their representation in livestock cooperatives and organizations.
10. **Strengthen Institutional Support**: Develop and implement policies and programs to provide institutional support for women in livestock, including access to credit, extension services, market information, and veterinary care tailored to their needs.

Examples of women farmers success stories- women empowerment

The Mulukanoor Women Cooperative Dairy (MWCD) is India's first women's cooperative dairy which is running successfully since 2002 with 21,000 women members (as on 2013) who are predominantly small holder dairy producers, and the dairy is now handling around 65,000 litres of milk per day. It is in Mulukanoor Village, Bhimadevarapalli Mandal, Karimnagar district, Telangana.

The Mulukanoor Women's Dairy Cooperative implemented various strategies to empower women members and overcome gender disadvantages.

These initiatives included technical training for women farmers to enhance their knowledge of livestock care, board membership opportunities for women, and direct payment of dairy income to women.

Gender Inclusivity in Milk Procurement Channels has been observed as both genders are involved in supplying milk in various dairies.

Cooperative is actively involving and promoting women in milk marketing activities.

Market information is shared predominantly among women in the existing procurement areas. This indicates that MWDC is ensuring that women have access to crucial market information, empowering them to make informed decisions.

The Theni Goat Farmer Producer Company Limited

Theni Goat Farmer Producer Company Limited an FPO in Tamil Nadu, stands out for its resilience and success during challenging times, particularly the COVID-19 pandemic. Led by women, this FPO has demonstrated the importance of community-driven development in agriculture. The success story began with the formation of an NGO called Vidiyal in 1994, which promoted savings and credit among women. In 2000, 250 Self Help Groups (SHGs) pooled their savings to form a corpus, leading to the creation of Vanavil. A needs-based assessment in 2007 identified goat rearing as a common need in the community. The NGO initiated capacity building and scientific training for goat rearing through innovative audio files accessible via mobile phones. This training continued for five years before the formation of the Farmer Producer Company (FPC) in 2015.The Theni Goat Farmer Producer Company Limited, with 1050 women farmers, achieved success by accessing markets, negotiating prices, and improving production processes. The FPO's profit-making and dividend distribution have been consistent for the past four years. To ensure sustainability, a mobile-based Massive Open Online Course (Mobi MOOC) on corporate literacy was conducted for illiterate shareholders.

During the pandemic, the FPO diversified its offerings, selling vegetables and necessities in the area. The success of the Theni Goat Farmer Producer Company highlights the potential of community-driven development and the impact of agriculture on women's empowerment. The women's dedication, training in goat rearing and financial management, and access to innovative technologies have elevated them as skilled entrepreneurs.

In summary, achieving women's empowerment in rural livestock production requires a comprehensive approach addressing capacity-building, resource access, cultural norms, and involving men as partners. Implementing these strategies empowers women to actively contribute to livestock production and make a meaningful impact on their households and communities.

Role of women in poultry farming

Women play a significant role in poultry farming(backyard poultry), particularly in rural communities where poultry raising is a crucial livestock activity. This sector offers various benefits to rural families, such as income generation, poverty alleviation, and improved nutrition through the production of valuable protein sources like meat and eggs. What sets poultry farming apart is that, unlike other livestock production systems, rural women often have control over the entire process, from feeding the birds to marketing the products.

For women who stay at home, poultry farming provides several advantages:

i. **Emergency support**: It offers a means to help the family during times of need, providing cash for emergencies.

ii. **Savings**: Women can save money for future investments, contributing to the household's financial stability.

iii. **Income for Family Needs**: Poultry farming allows women to generate income to meet their children's and household's needs.

iv. **Nutritional Support**: It supplements the family's protein intake, contributing to better nutrition.

Poultry farming also presents various advantages:

i. **Ease of Management**: Poultry are relatively easy to manage, requiring few external inputs, and they enjoy good market demand and prices.

ii. **Resource Recycling**: Feeding backyard poultry involves recycling household and farm waste, making efficient use of resources.

iii. **Innovation**: Women often come up with innovative ways to utilize waste products in poultry farming.

iv. **Poverty Reduction**: Poultry farming can be a powerful tool for poverty reduction among rural women and children.

Moreover, by increasing women's income, poultry farming enhances their social status and decision-making power within the household.

However, women involved in poultry production face several constraints:

i. **Market Challenges:** They often deal with issues like distance from markets, limited access to market information, and inadequate transport facilities.

ii. **Healthcare Access:** Limited access to vaccines and veterinary services.

iii. **Training and extension services:** Women may have limited access to training and extension services. Although extension helps in providing information and promote adoption of technologies, provision of extension services in developing countries remains low for both men and women, and women tend to make less use than men of extension services (Meinzen-Dick *et al.*, 2010).

iv. **Credit Access:** Limited access to credit can result in insufficient funds for quality feed and medicines.

v. **Changing Roles:** Traditionally viewed as a women-dominated domain, poultry keeping is experiencing a shift in roles and attitudes, as men recognize its economic value, potentially affecting women's control over the generated income. Women's participation in backyard poultry is decreasing due to commercialisation of poultry The contribution of backyard poultry to the total egg and chicken production is decreasing over the years due to fast growing commercial farms.

To address these challenges and support women in poultry production:

i. **Marketing**: Establish well-organized marketing systems accessible to women to ensure better prices for poultry products.

ii. **Training**: Provide training in poultry management practices and improve access to poultry health services to enhance the success of poultry activities.

iii. **Access to Credit**: Ensure availability of credit, which is often essential for poultry development in rural areas.

iv. **Gender Integration**: Incorporate gender considerations into poultry projects to identify factors of production and access to benefits.

v. **Policy Support**: Advocate for policies that recognize and support the crucial role women play in poultry production and rural development, involving policymakers and planners in these discussions.

vi. Encouraging and formation of women's groups will go a long way in empowering them

A Comparative Analysis of Urban and Rural Women's Participation in Livestock production

S.No.	Aspects	Urban rich women	Rural poor women
1	Species of animals reared	Bovines, Backyard poultry	Bovines, Ovines and Backyard poultry
2	No. of animals	Large	Small
3	Direct Participation in livestock production activities (grazing, collection of grasses, feeding, cleaning of sheds and milking of animals	Low (Get the work done through labour and hence more of supervisory role. Restrict themselves within the house/farm	High, direct involvement in all the activities
4	Access to and control over resources	High	Low
5	Participation in Decision making i. Sale and purchase of animals ii. Household consumption of livestock products, sale of livestock products	Low in case of bovines, but high in backyard poultry High	Low in case of bovines and high in case of ovines and backyard poultry Low in case of milk but high in case of eggs
6	Access and control over livestock income	High	Low
7	% Contribution to family income	Low	High

Women-centric value chain development in Bangladesh

The Microfinance and Technical Support Project (MTFSP) is an IFAD-supported project in Bangladesh. This project started in 2004 to improve livelihoods and food security of moderately poor and very poor households and the empowerment of women, through the promotion of sustainable income-generating activities and livestock technologies.

Under the MFTSP, IFAD has supported the development of women-centred value chains at the village level. In the MFTSP approach for the poultry value chain, the activities of a single woman in backyard poultry production are disaggregated into a set of clearly distinguishable activities, which are then carried out by several women.

Occupations are created for each activity, and specialized training is delivered to each of the actors: model poultry breeders, mini hatchery owners, chicken rearers, and poultry keepers. Value has been added through (a) commercializing the transactions between each node, and (b) improving the genetic material, thus raising overall income levels for each actor.

In this way, a female income stream has been generated for the household. This raises overall household income in households where every Taka (basic unit of Bangladeshi currency) counts. Consensus exists that although social norms dictate that men are responsible for supporting the family economically, poverty levels have meant that women are frequently seen as a burden – literally an extra mouth to feed.

Assisting women to earn money has brought about more equitable roles and relations in the household, recognition of women's contribution to the household economy, and generally an important increase in the status of women, both within the household and indeed within the village. Apart from the final interface with the market, the transactor between value chain actors is female. Mini-hatchery owners buy fertile eggs from other project participants who run small parent farms in confined production conditions.

The parent lines are Fayoumi females and Rhode Island Red males. After hatching, the mini hatchery sells day-old Sonali chicks to chick rearing units. After 8 weeks, the chick rearers sell the young birds to another category of project participants, poultry keepers, who keep poultry for egg production and for sale to the market.

The MFTSP has been successful in targeting women because it has created a value chain that is geographically limited. It trains women to be specialized actors at well-defined nodes in the chain and adds value by upgrading and managing gene flow. The level of technology is appropriate for women. The mini hatcheries are easy to build and manage, and yet are sophisticated in design.

Further success factors include the fact that the project specifically set out to reach women and the poorest, and it has benefited from committed staff both within government departments and in the implementing organizations.

The Madhya Pradesh Women's Poultry Producer Company Limited (MPWPCL)

The Madhya Pradesh Women's Poultry Producer Company Limited (MPWPCL) stands as a remarkable example of women empowerment and sustainable agricultural practices in the state. Established in 2002, this federation of ten poultry producer cooperatives has evolved into a significant player in the poultry industry, particularly in broiler chicken production.

With over 5,000 women members hailing from low-income tribal communities, MPWPCL not only facilitates livelihood creation but also addresses the financial challenges faced by its members. The cooperative provides comprehensive support, including production assistance, inputs, advisory services, and marketing, empowering women in small-holder poultry farming.

One notable aspect of MPWPCL's success lies in its ability to adapt industrial poultry practices to suit the needs of small women farmers in remote villages. By organizing women into collectives and implementing effective systems and processes, the cooperative has achieved industry-competitive production and scale efficiencies.

The cooperative's operations include a hatchery, feed plant, and pallet plant, contributing to the overall sustainability of its activities. Currently serving around 10,000 members, all of whom are economically disadvantaged women managing flocks ranging from 400 to 800 birds, MPWPCL plays a crucial role in providing a regular income to its members.

In addition to the economic benefits, MPWPCL assists its members in accessing loans, preventing them from falling into the debt trap. The cooperative has successfully created its market brand, "Sukhtava Chicken," focusing on broiler meat and eggs. This brand is now operational in approximately 12 cities across Madhya Pradesh.

The retail outlets of Sukhtava Chicken are a testament to modern and hygienic poultry processing practices. These outlets incorporate state-of-the-art tools and techniques, offering consumers a complete package of value for money. The success of MPWPCL is evident in its impressive financial figures, with a turnover of 290 crores and profits ranging between 60-70 crores.

The growth trajectory of MPWPCL is nothing short of remarkable. Starting with a turnover of only `15 lakh in 2007, the cooperative has grown exponentially, demonstrating a 231.82% Compound Annual Growth Rate (CAGR). This unprecedented growth can be attributed to a robust marketing strategy and the unwavering commitment of poor women involved in the cooperative.

The company follows a unique model based on the 4 Ps - Products, Price, Promotion, and Place, which has contributed significantly to its success. Over the past ten years, the membership has surged from an initial 60 to an impressive 10,000, reflecting a CAGR of 176.55%. With a clear vision for the future, MPWPCL aims to further expand its membership to 20,000 in the next five years, solidifying its status as a fast-growing company in terms of both turnover and membership. The cooperative's success story is not just about economic growth but also about transforming the lives of women in tribal communities, proving that sustainable agriculture can be a powerful tool for empowerment and upliftment.

"The Daily Toil: Women's Unseen Struggles in Animal Care"

Every morning, as the first rays of sunlight gently touch the horizon, a dedicated group of women in rural communities across Haryana and West Bengal begins their day. Their work is not confined to the cozy comforts of modern machinery or the convenient push of a button; instead, it's a laborious, hands-on engagement with the animals that form the backbone of their families' livelihoods.

The tasks these women undertake are far from trivial. They encompass a wide array of physical chores, from scrubbing and cleaning to nourishing and tending to the needs of their livestock. Each day brings the same unyielding routine: the women rise early, their commitment unwavering. They commence by tidying up the animal sheds, collecting dung with the unwavering dedication that characterizes their lives.

Once the chores within the confines of the shed are complete, they turn their attention to the well-being of their animals. Carefully, they feed them, nurturing their health and vitality. The animals are their livelihood, and their welfare is paramount.

But the responsibilities do not end there. With a practiced hand and a heart filled with determination, these women take on the task of milking their animals, providing a valuable source of nourishment for their families. It is a delicate dance of skill and precision, a daily routine they execute with grace.

As the sun continues its ascent into the sky, it is time to lead the animals to graze, allowing them to roam the open fields and pastures. The women, tireless and relentless, watch over their charges as they munch on the grasses, ensuring their safety and well-being.

This routine repeats itself in the evening, with the same unwavering dedication and attention to detail. Animal care knows no respite. It is a year-round commitment that demands daily vigilance.

Surveys conducted in these rural regions reveal a striking truth: on average, women invest approximately two hours of their day in these demanding animal care activities. Two hours might seem fleeting, but when accumulated over a year, it amounts to a staggering 730 hours of relentless dedication. To put it in perspective, this is the equivalent of around 104 full days of continuous labor. The average female labour participation in various dairy farming activities was quite substantial and it was reported to be about 33 per cent (Dhaka *et al.*, 1993).

This dedication, this relentless toil, defines these women as usual-status principal workers. Any woman who reports more than 100 days of this labor in a year has not merely adopted a role; she has become the backbone of her family's sustenance. Her unwavering commitment to the animals transcends the boundaries of routine; it exemplifies the true essence of resilience and devotion, silently weaving the fabric of her family's survival.

Summary

- Women play a pivotal role in the livestock sector in India, contributing significantly to the welfare of rural communities and the nation's economy.
- The livestock sector is a lifeline for the rural population, with approximately 8.80% of the total population employed in this sector.
- Women in India are the primary workforce in animal husbandry, with various statistics underscoring their essential contributions.
- An estimated 400 million impoverished livestock keepers globally are women, out of 600 million in total.
- Women make up around 63% of the workforce in the agriculture sector in India.
- In rural agriculture, 11% of female workers are employed in livestock raising.
- India's Employment and Unemployment Survey (EUS) in 2012 reported around 8 million female and 2 million male workers in rural livestock raising.

- Women dedicate around 3 to 5 hours a day to livestock production activities, representing a substantial time commitment.
- In India, an impressive 75 million women are engaged in dairying, compared to 15 million men.
- The dairy sector employs 70% of its workforce as women.
- The number of women members in dairy cooperatives increased from 5 million in 2015-16 to 5.4 million in 2020-21.
- Women now account for 31% of all members of dairy producer cooperatives in India.
- Women actively participate in various aspects of livestock management, including feeding, general management, health and breed management, processing, and marketing.
- Despite their vital roles, women face numerous challenges in livestock production, such as limited access to resources, gender biases, and cultural norms.
- Recognizing and supporting women's contributions in animal husbandry is essential for enhancing the sustainability of rural communities and the well-being of the animals they care for.

SECTION B: DESCRIPTIVE QUESTIONS

1. How does livestock ownership contribute to women's empowerment in rural India?

Livestock ownership contributes to women's empowerment in rural India in several ways. First, it provides women with an opportunity to engage in income-generating activities and become financially independent. Livestock production is a preferred activity for women in many women self-help groups and many government programs aimed at women empowerment. Women comprise a considerable proportion of the workforce in livestock production, and their engagement in livestock production leads to higher income from livestock. In fact, households with more illiterate female workers realize more income from livestock compared to households with illiterate male workers. This increased income gives women more control over resources within the household, enhancing their participation in decision-making processes.

Second, livestock ownership allows women to have a say in resource allocation within the household. Women who have control over income from livestock production are more likely to spend it on nutrition, health, and education of their children. Studies have shown that when women have a greater contribution to household income, it leads to better nutritional outcomes for children.

Livestock ownership can thus provide women with the means to improve the well-being of their families and invest in their children's future.

Third, livestock ownership can lead to improved access to nutritious food and healthcare for women and their families. Livestock provide a source of complete protein, energy, and micronutrients that are often lacking in rural diets. The consumption of animal source foods (ASFs) from livestock can significantly impact child nutrition, especially for children between 6 and 24 months of age. Additionally, the income generated from livestock can be used to purchase nutritious food items and better healthcare facilities, improving the overall nutrient availability for family members.

In conclusion, livestock ownership plays a critical role in women's empowerment in rural India. It provides women with income-generating opportunities, enhances their control over resources, and improves the well-being of their families. Livestock ownership can contribute to better nutrition, healthcare, and education outcomes for women and their children, leading to their empowerment and improved livelihoods.

2. How can livestock serve as a source of income for rural households, particularly for women, and what role do gendered interventions play in enhancing women's skills in livestock production and marketing?

1. **Income generation**: Livestock can serve as a source of income for rural households, especially for women who are often involved in livestock rearing and management. Gendered interventions can provide women with training, resources, and support to enhance their skills in livestock production and marketing. This can lead to increased incomes for women and their households, contributing to poverty reduction and economic empowerment.
2. **Livelihood opportunities**: Livestock-related interventions can create additional livelihood opportunities for women in rural areas. By promoting women's engagement in livestock activities, such as animal care, breeding, and value-added product development, gendered interventions can help women diversify their income sources and improve their economic resilience.
3. **Empowerment and decision-making**: Gendered interventions can also focus on empowering women by enhancing their control over resources, including income generated from livestock activities. This can be achieved through capacity-building programs, promoting women's participation in decision-making processes related to livestock, and advocating for women's property rights

4. **Improved household well-being**: Livestock ownership and the income generated from it can have a positive impact on various aspects of household well-being, including health, nutrition, and education. Gendered interventions that prioritize women's involvement in livestock activities can contribute to better health and nutrition outcomes for both women and children. Additionally, increased income from livestock can enable households to invest in education and healthcare, leading to improved overall well-being.

   In summary, gendered interventions in livestock-related activities can improve families diet quality and income in rural India by promoting the production and consumption of ASFs, creating livelihood opportunities for women, empowering women, and enhancing overall household well-being. These interventions can have a transformative effect on rural communities, contributing to poverty reduction, gender equality, and sustainable development.

3. What is the role of women in livestock production in India?

The role of women in livestock production in India is widely acknowledged. They contribute significantly to various aspects of livestock rearing and management. In fact, in India, more than three-fourths of the labour requirement for livestock production is met by women. Women are primarily responsible for tasks such as animal care, fodder collection, chaffing, feeding, milking, and other day-to-day activities related to livestock. They play a crucial role in ensuring the health and well-being of the animals. Furthermore, women's involvement in livestock production has the potential to enhance their empowerment and improve their socio-economic status. Livestock ownership provides women with a means to generate income and gain control over resources within the household. It allows them to have a say in decision-making processes and resource allocation, which can lead to improved household welfare and better nutrition, health, and education outcomes for their children. Livestock production also offers women opportunities for skill development and entrepreneurship. For instance, women are actively involved in dairy cooperatives, where they comprise 28% of the total membership. These cooperatives provide women with a platform to engage in income-generating activities, gain access to markets, and enhance their economic independence. The NDDB is promoting women dairy cooperatives to enhance women's participation in cooperative system.

Overall, women's participation in livestock production in India is vital for ensuring food security, poverty reduction, and gender equality. Recognizing and promoting their role in this sector can lead to enhanced livelihood opportunities, improved household well-being, and greater gender empowerment.

4. Are there any successful examples of programs or initiatives that have effectively empowered women in agriculture& animal husbandry sector?

Yes, there are several successful examples of programs and initiatives that have effectively empowered women in agriculture. For instance, the *MahilaKisanSashaktikaranPariyojana* (MKSP) is a government-led initiative that aims to empower women in agriculture by providing them with training, information, and resources to improve their productivity and income. Similarly, the Self-Employed Women's Association (SEWA) is a non-governmental organization that has been successful in organizing women farmers and providing them with access to credit, markets, and other resources. Another example is the Women in Agriculture (WIA) program, which is a joint initiative of the International Fund for Agricultural Development (IFAD) and the Government of India that aims to empower women in agriculture by providing them with training, credit, and other support services. These programs have been successful in improving women's access to resources, increasing their productivity and income, and enhancing their decision-making power in agriculture. Similarly, A-HELP (Accredited agent for Health and Extension of Livestock Production) is an initiative that involves community-based women activists who bridge the gap between local veterinary services and livestock owners while providing primary services. The goal of A-HELP is to improve livestock health and productivity, as well as to empower women in the dairy sector.

5. How can livestock research and development programs be designed to address social inequalities and promote transformative change in gender relations within agricultural communities?

Livestock research and development programs can be designed to address social inequalities and promote transformative change in gender relations within agricultural communities by adopting a gender-transformative approach. This approach involves integrating gender considerations throughout the program design and implementation process.

Firstly, it is important to go beyond simply including women in the program and to consider how to improve the position of women and influence strategic gender relations. This means moving away from generic beneficiary groups and establishing partnerships with transformative development programs that enable women to benefit from agricultural innovations.

Secondly, a gender-transformative approach recognizes that gender inequalities are deeply rooted in social norms, attitudes, behaviours, power relations, and social systems. Therefore, it is essential to address these underlying factors that perpetuate gender inequalities. This includes engaging with the political dimensions of women's empowerment and acknowledging that intensive efforts and resources will be required to achieve change.

Furthermore, livestock research and development programs should aim to enhance women's access to productive resources, such as land, water, and credit, which are often limited for women. They should also focus on improving women's access to market information and prices, as well as increasing their decision-making powers within households and communities

In addition, livestock programs can contribute to reducing gender disparities in asset ownership by recognizing that livestock are assets that women can more easily own. Interventions that increase women's access and rights to livestock, and safeguard their ownership from dispossession and theft, can empower women and help them move out of poverty.A study by van Eerdewijk and Danielsen (2015) concluded that if gender relations and dynamics (access to and control over resources), as well as values and assumptions, are not considered, it is unlikely that there will be an impact on gender relations and women's position.

Overall, a gender-transformative approach in livestock research and development programs involves addressing social inequalities, challenging gender norms, and empowering women within agricultural communities. By integrating gender considerations throughout the program design and implementation process, these programs can contribute to more equitable development outcomes for men, women, households, and communities.

6. Discuss certain gender outcomes in dairy farming?

1. Women's ownership of dairy cows is positively associated with their productive labour time but may be negatively associated with their leisure time and social participation.
2. Women who own dairy cows spend more time on care work than women who do not own cows, which may limit their ability to engage in other activities.
3. Women who own dairy cows have higher levels of decision-making power and control over household resources, which can contribute to their empowerment.
4. The benefits of dairy livestock assets for women's empowerment are not uniform across diverse groups of women, and depend on factors such as age, marital status, and household composition.

   Dairy livestock assets can have positive effects on women's productive labour time and decision-making power, high esteem in the community, better credit worthiness etc. They can also have negative effects on their leisure time and social participation. Therefore, importance of recognizing women's heterogeneity and agency in agricultural interventions should be studies for more nuanced and gender-sensitive approaches to development programming.

7. How does the ownership of dairy cows affect women's time poverty, and what are the implications for household food and nutrition security?

The ownership of dairy cows can affect women's time poverty in several ways. Owning dairy cows can increase women's productive labour time, as they are responsible for management of the cows and preparing milk products. This can lead to an increase in income and food security for the household, as dairy products can be sold or consumed by the family. But owning dairy cows can also increase women's time poverty, as they may have less time for other activities such as care work, income-generating activities, and social participation. This can lead to a decrease in their overall well-being and quality of life.

The implications for household food and nutrition security depend on how thc benefits of dairy cattle are distributed within the household. If women have control over the income and resources generated by the cows, they may be able to use them to improve the nutritional status of their families. However, if men or other household members control the income and resources, women may not have the power to make decisions about how they are used. This can lead to a situation where women are responsible for the care and feeding of the cows(inputs), but do not benefit from the income or resources(outputs) generated by them. In this case, the benefits of dairy animals may not translate into improved food and nutrition security for the household.

8. What are some of the challenges and opportunities for promoting women's empowerment and gender equality in livestock management, and how can development institutions address them effectively?

The challenges and opportunities for promoting women's empowerment and gender equality in livestock production are complex and context specific. Some of the key challenges include:

1. Gender norms and stereotypes that limit women's access to resources, information, and decision-making power.
2. Lack of access to education and training opportunities for women, which can limit their ability to engage in productive activities and make informed decisions.

3. Limited access to credit and financial services, which can prevent women from investing in their farms and businesses.
4. Inadequate infrastructure such as roads, markets, and extension services, which can limit women's ability to access inputs, sell their products, and obtain information and support.

At the same time, there are also opportunities for promoting women's empowerment and gender equality in livestock production, such as:

1. Increasing women's access to land, credit, and other productive resources.
2. Providing education and training opportunities for women, including in areas such as fiscal management, marketing, and leadership.
3. Strengthening women's participation in decision-making processes at the household, community, and national levels.
4. Investing in infrastructure and services that benefit women, such as rural roads, markets, and extension services.

To address these challenges and opportunities effectively, development institutions can take several steps, such as:

1. Adopting a gender-sensitive approach to programming, which considers the different needs, priorities, and constraints of women and men.
2. Engaging with local communities and stakeholders to understand the social, cultural, and economic context in which women live and work.
3. Building partnerships with local organizations and institutions that have experience and expertise in promoting women's empowerment and gender equality.
4. Monitoring and evaluating programs and interventions to ensure that they are achieving their intended outcomes and are responsive to the needs of women and men.
5. Setting up quotas for women or involving members' spouses can boost women's effective participation in and leadership of producer organizations (POs). Organizing seminars for women and facilitated by female extension advisers, talking about problems that women face in their activities, can also be a valuable strategy (Petrics *et al.*, 2015).

By taking these steps, development institutions can help to promote women's empowerment and gender equality in livestock production and contribute to more inclusive and sustainable development outcomes.

9. Discuss the role of Operation Flood in Women Empowerment in India?

Operation Flood, launched in 1970, brought about changes in women's work in raising animals, particularly in the states of Gujarat, Rajasthan, Karnataka, and Maharashtra in India. The program increased female family work across different social classes and castes. Women played a significant role in tending to animals, cleaning and disposing of dung, feeding and milking of cows, selling milk to dairy coops. Supply of milk to dairy coops helped the dairy farmers in getting money at regular intervals in a sustainable manner and this income was used by the women to meet the essential requirements of the family.

Operation Flood also increased women's participation in raising animals due to increased incomes, more employment opportunities, and the establishment of dairy cooperatives. Women also took on the responsibility of caring for animals during crop failures in some states when men migrated out of distress conditions for extended periods.

The program highlighted the importance of women's participation in raising dairy cattle and provided economic opportunities for rural women. Overall, Operation Flood played a role in empowering women by increasing their involvement in livestock raising and providing them with economic opportunities in rural areas.

10. What is the level of self-employment among women livestock farmers, and how does the presence of casual labourers in livestock raising compare?

It is evident that most women livestock farmers around 95%, are self-employed in the livestock sector. This highlights a significant level of entrepreneurship and self-reliance among rural women in this field. Moreover, there seems to be a limited presence of casual workers in livestock raising, indicating that the work is primarily managed by those who own the livestock or are directly responsible for their care. There are millions of families who are landless but own dairy cattle to sustain their livelihoods and, in these families, women play a significant role.

The engagement of rural women in multiple activities based on seasonal variations and the availability of work is a common and adaptive practice. Livestock raising is just one of these activities, and it is often conducted within the homestead by household members. This integration of livestock management into the household setting demonstrates the versatility and resourcefulness of rural women in balancing various responsibilities and leveraging their resources to sustain their livelihoods.

11. What are the factors that contribute to higher participation of women in livestock raising in the Northern States?

The higher participation of women in livestock raising in the Northern States of India can be attributed to several factors:

i. **Concentration of milch animal-owning households:** According to a study conducted by NDDB, Northern States like Rajasthan and Uttar Pradesh have a higher concentration of households that own milch animals. This indicates a higher prevalence of dairy farming in these states, which requires women's participation in raising bovines.

ii. **Ownership of bovines:** The Land and Livestock Holdings survey found that Uttar Pradesh and Rajasthan together owned around 40% of the total bovine holdings in India. As these states have a higher ownership of bovines, women's participation in livestock raising, particularly in raising bovines, is naturally higher.

iii. **Cultural and traditional practices:** Livestock raising, including raising bovines, has been a part of the cultural and traditional practices in the top five milk producing States of India (Uttar Pradesh, Rajasthan, Madhya Pradesh, Gujarat and Andhra Pradesh). Women have traditionally been involved in activities related to livestock rearing, and these practices have continued over the years.

iv. **Economic importance:** Livestock raising, especially dairy farming, holds significant economic importance in the Northern States. Women's participation in dairy cattle raising contributes to household income and livelihoods, making it a viable and important economic activity for them.

It is important to note that these factors are specific to the Northern States of India and may not necessarily apply to other regions.

12. How does the oversight of gender dynamics in animal husbandry, particularly the exclusive focus on male decision-makers, impact the success of development projects, and what are the specific roles that women play in ensuring the comprehensive care and management of animals within households?

Gender should not be viewed in isolation; it permeates all facets of research and development where socio-economic and power dynamics are at play. In the planning and execution of agricultural development initiatives, the household is commonly treated as the analytical unit, with male heads of households seen as the primary decision-makers. Unfortunately, this approach often side-lines the roles of other household members. This oversight can impede project success, given that women, men, and children bring distinct skills, resources, priorities, and responsibilities to farm production. In animal husbandry, for instance, if we adopt a similar mind-set, the exclusive focus on male decision-makers might neglect the crucial roles that women play. Women may possess specific expertise, priorities, and responsibilities in the care and management of animals. For example, while men may handle tasks like constructing animal shelters, women might be more involved in day-to-day animal care, ensuring proper feeding, hygiene, and health monitoring. Disregarding these roles can undermine the effectiveness of animal husbandry projects and limit the overall success of the initiatives. Therefore, a more inclusive approach that recognizes and integrates the diverse contributions of all household members is essential for comprehensive and impactful development in animal husbandry.

Gender analysis is a useful tool to identify the gender division of labour, the demand for advisory services from men and women, including the crops and the roles associated with men and women, and the ways in which this demand could be satisfied. Women can also be targeted separately, especially if the crop or activity concerned is considered a women's task (e.g. backyard poultry, goat-rearing, and home-garden vegetable cultivation), and particularly in highly segregated society (Carter and Weigel, 2011).

SECTION C: SHORT ANALYTICAL QUESTIONS

1. **How do women's contributions to livestock management impact rural communities in India?**

   Women's contributions to livestock management in India play a crucial role in sustaining rural communities. They ensure the well-being of livestock, generate income, and provide essential resources for their families. Their active involvement in feeding, healthcare, and processing activities enhances food security and livelihoods, making them the backbone of rural economies.

2. **What challenges do women in India face in the livestock sector, particularly in financial endeavours?**

   Women in the Indian livestock sector face challenges in participating in financial endeavours, such as securing loans and managing business records. Limited access to financial resources hinders their economic empowerment. Despite their significant role in other aspects of livestock management, addressing these challenges is vital for their holistic empowerment.

3. **How does livestock ownership empower women in India, and what are the broader impacts on households and communities?**

   Livestock ownership empowers women in India by enhancing their decision-making power, improving household welfare, generating income, and boosting self-esteem. This empowerment has a cascading impact on households and communities, leading to more equitable decision-making, improved food security, poverty reduction, and inclusive and sustainable development.

4. **What is the significance of women's participation in dairy cooperatives in India's animal husbandry sector?**

   Women's participation in dairy cooperatives is significant as it allows them to collectively market their dairy products, gain access to resources, and strengthen their bargaining power. It also fosters economic independence and self-reliance among women in rural areas, contributing to their empowerment and overall community development.

5. **How does the role of women in livestock management vary across different states in India, and what factors influence these variations?**

   The role of women in livestock management varies across different states in India, with some states having a higher percentage of women engaged in these activities. This variation can be influenced by factors such as cultural traditions, economic conditions, and the prominence of the livestock sector in a particular region.

6. **What are the health and well-being implications for women engaged in livestock activities, particularly in terms of musculoskeletal disorders?**

   Women engaged in livestock activities often experience musculoskeletal disorders (MSD), including back pain. This has health and well-being implications for women, affecting their physical health and potentially reducing their ability to carry out these tasks effectively. Addressing the health concerns of women in the livestock sector is important for their overall well-being.

7. **How can initiatives and policies be designed to further empower women in the Indian animal husbandry sector, especially in terms of financial access and decision-making roles?**

   Initiatives and policies aimed at empowering women in the Indian animal husbandry sector should focus on increasing their access to financial resources, providing training and education, and promoting their active participation in decision-making roles. By addressing these areas, women can be better integrated into the sector and contribute more effectively to their families and communities.

8. **What are the long-term economic and social benefits of empowering women in animal husbandry for rural India?**

   Empowering women in animal husbandry leads to long-term economic benefits, including increased income, poverty reduction, and improved food security. Additionally, it has social benefits, such as enhanced self-esteem, more equitable decision-making within households, and better overall well-being for rural communities, contributing to sustainable development.

9. **What are the key obstacles and constraints faced by women in livestock production?**

   Women in livestock production face challenges such as insecure land tenure, limited access to resources and credit, gender roles and labor division, and gender biases in access to extension services and decision-making power.

10. **How does insecure land tenure affect women's participation in livestock production?**

    Insecure land tenure hinders women's ability to access and own land for livestock production, limiting their engagement in the sector and their economic empowerment.

11. **Why is the intensification of livestock production a potential obstacle for women?**

    The shift towards intensive commercial livestock production systems may marginalize women's participation as these systems often require significant capital and resources that women may have limited access to.

12. **What measures can be taken to address the limited access to resources by women in livestock production?**

    Measures include providing women with access to credit, extension services, and training, as well as promoting ownership and control of livestock assets.

13. **How can cultural norms and gender biases be challenged to empower women in livestock production?**

    Cultural norms and biases can be challenged through awareness campaigns, sensitization programs, and advocacy efforts that promote gender equality, and through the involvement of men as allies in the process.

14. **Why is women's participation in poultry farming particularly important in rural communities?**

    Women's participation in poultry farming is crucial in rural communities because it can provide income, nutritional support, and financial stability to households, contributing to poverty alleviation and improved nutrition.

15. **What are some of the advantages of poultry farming for women in rural areas?**

    Poultry farming provides emergency support, savings, income for family needs, and nutritional support, and it is relatively easy to manage, making efficient use of resources.

16. **What challenges do women in poultry farming face, and how can these challenges be addressed?**

    Challenges include market issues, limited access to healthcare, knowledge and technology, and credit access. They can be addressed through improved marketing systems, training, access to credit, and gender integration in poultry projects.

17. **How can the involvement of men in poultry farming impact women's roles and control over generated income?**

    The increasing involvement of men in poultry farming may impact women's traditional control over income, potentially leading to shifts in roles and attitudes, which must be carefully managed to ensure women's empowerment.

18. **What role can policy support play in promoting women's participation in livestock production?**

    Policy support can recognize and support the vital role women play in livestock production and rural development, ensuring that policies incorporate gender considerations and involve policymakers and planners in these discussions.

19. **How does inclusive livestock development contribute to the achievement of SDG5 (gender equality & empowerment of women)?**

    Inclusive livestock development is identified as a catalyst for achieving SDG5. The data suggests that empowering women in livestock farming can significantly contribute to gender equality and women's empowerment, particularly in rural areas. Understanding the role of livestock in broader sustainable development goals is crucial for formulating effective policies and interventions.

20. **According to the Food and Agriculture Organization (FAO) report, what are the potential benefits of ensuring gender equality in livestock farming?**

    The FAO report suggests that if women have access to the same level of resources as men, agricultural productivity could increase by 10–30%, and agricultural output could rise by up to 4%. This emphasizes the significant positive impact that gender equality can have on agricultural productivity and output.

21. **Considering the challenges faced by women in livestock activities, what measures can be taken to address gender disparities and improve the well-being of female livestock workers?**

    Possible measures include implementing ergonomic practices to reduce musculoskeletal disorders, providing training and education to enhance women's skills in financial endeavours, and promoting equal access to resources and opportunities. Additionally, recognizing and valuing women's diverse roles in livestock management is crucial for fostering a more inclusive and supportive environment.

22. What are some key statistics that highlight the significant contributions of women in India's animal husbandry sector?

Several statistics underscore the pivotal role of women in agriculture and livestock production in India. For instance, nearly 63% of workers in the agriculture sector are female, and around 11% of the rural female agricultural workforce is employed in livestock raising. Moreover, the dairy sector employs 70% of its workforce as women, with a substantial increase in the number of women's dairy cooperative societies.

23. What is the underestimated role of women in animal husbandry, and why is it important to recognize their contributions?

The underestimated role of women in animal husbandry involves their significant contributions to animal care on farms. Recognizing these contributions is crucial because official surveys often underreport the number of women involved in this work, and acknowledging their role is essential for promoting gender equity in livestock production.

24. What are the measures suggested to promote women's empowerment in livestock production?

Measures include ensuring secure land tenure, enhancing access to capital and knowledge, promoting ownership and control of livestock assets, addressing gender roles and labor division, implementing capacity-building programs, strengthening women's networks and organizations, promoting gender equality, engaging men in the process, enhancing access to resources, and challenging cultural and social norms.

25. Why has there been a decline in women's participation in livestock raising in India since 2004-05, and what factors contribute to this decline?

Factors contributing to the decline include changing preferences, decline in cattle and buffalo ownership, and regional variation. Younger women may prefer alternative income-generating activities, reduced ownership of livestock may decrease participation, and regional variations may impact women's engagement in raising animals.

26. What strategies are suggested to promote women's empowerment in decision-making activities related to livestock production?

Strategies include ensuring secure land tenure, enhancing access to capital and knowledge, promoting ownership and control of livestock assets, addressing gender roles and labor division, implementing capacity-building programs, strengthening women's networks and organizations, promoting gender equality, engaging men in the process, enhancing access to resources, and challenging cultural and social norms.

27. What are the advantages and disadvantages faced by women in poultry farming, particularly in rural communities?

Women in poultry farming have opportunities for emergency support, savings, income generation, and nutritional support. Poultry farming is relatively easy to manage, allows for resource recycling, encourages innovation, and contributes to poverty reduction. Challenges include market issues, limited access to healthcare and technology, and changing gender roles within the industry. Solutions involve addressing market challenges, providing training, improving healthcare access, ensuring credit availability, and promoting gender integration in poultry projects.

28. How does poultry farming impact the social status and decision-making power of women in rural areas?

Poultry farming enhances the social status and decision-making power of women by increasing their income. With control over the entire process, from feeding to marketing, women gain economic independence. This economic empowerment contributes to their elevated position within the household and community.

29. In what ways do women contribute to animal care in rural communities, and what challenges do they face in this role?

Women contribute significantly to animal care by engaging in tasks such as cleaning, feeding, milking, and ensuring the overall well-being of livestock. They play a crucial role in maintaining the livelihoods of their families. Challenges include market issues, limited healthcare access, knowledge and technology barriers, restricted credit access, and changing gender roles within the animal care domain. Solutions involve addressing market challenges, providing training, improving healthcare access, ensuring credit availability, and promoting gender integration.

30. How does the daily toil of women in animal care impact their social and economic standing in rural communities?

The daily toil of women in animal care defines them as principal workers, showcasing their resilience and devotion. This labour-intensive role contributes significantly to their families' sustenance. Women invest approximately two hours a day, totalling 730 hours annually, in these activities, highlighting their essential role in rural economies.

31. What are the key statistics and contributions of women in the animal husbandry sector in India?

Women make up around 63% of the workforce in the agriculture sector in India. In rural agriculture, nearly 11% of female workers are employed in livestock raising.India's Employment and Unemployment Survey (EUS) in 2012 reported around 8 million female and 2 million male workers in rural livestock raising. Women dedicate around 3 to 5 hours a day to livestock production activities, representing a substantial time commitment. Women account for 31% of all members of dairy producer cooperatives in India.

32. What are the challenges faced by women in livestock production, and how can these challenges be addressed?

Women in livestock production face limited access to resources, gender biases, and cultural norms. To address these challenges, initiatives should focus on providing better access to resources, challenging gender biases, promoting cultural awareness, and creating supportive policies that recognize and value the contributions of women in livestock production.

33. How can policymakers and planners support women's roles in poultry farming and animal care in rural development?

Policymakers and planners can support women in poultry farming and animal care by:
Establishing well-organized marketing systems accessible to women.

Providing training in husbandry practices and improving access to poultry health services.

Ensuring availability of credit for poultry development in rural areas.

Incorporating gender considerations into poultry projects to identify factors of production and access to benefits.

Advocating for policies that recognize and support the crucial role women play in poultry production and rural development.

SECTION D: OBJECTIVE QUESTIONS

Question 1 Multiple Choice Questions

1. **On average, how many hours per year do female farmers in India dedicate to dairy activities?**

   a) 100 hours   b) 315 hours

   c) 1,000 hours   d) 1,779 hours

2. **What percentage of female farmers in livestock activities experience musculoskeletal disorders?**

   a) 10%   b) 25%

   c) 45%   d) 55.3%

3. **What is the primary source of income for many rural women in livestock farming?**
   a) Crop production b) Off-farm employment
   c) Poultry farming d) Dairy farming

4. **What advantages does livestock ownership offer to women?**
   a) Enhanced decision-making power b) Reduced workload
   c) Access to free resources d) Lower self-esteem

5. **How can livestock ownership contribute to food security for women?**
   a) By reducing the number of livestock owned
   b) By providing a consistent supply of nutritious animal-based foods
   c) By decreasing income generation
   d) By reducing women's role in decision-making

6. **How does income from livestock benefit households?**
   a) It leads to overconsumption.
   b) It contributes to overall well-being.
   c) It increases poverty.
   d) It has no impact on household welfare.

7. **What role does livestock play in income generation for women?**
   a) It has no impact on income.
   b) It reduces income.
   c) It serves as a reliable source of cash income.
   d) It decreases self-esteem.

8. **How does active involvement in livestock-related activities affect women's self-esteem?**
   a) It lowers self-esteem b) It has no impact on self-esteem
   c) It reduces decision-making power. d) It bolsters self-esteem.

9. **What type of assets are livestock compared to land or financial assets?**
   a) Less accessible b) More accessible
   c) Equally accessible d) Unimportant

10. **What is the importance of women's income management in the household?**
   a) It leads to domestic violence.
   b) It doesn't impact resource allocation.

c) It often leads to improved resource allocation.
d) It has no influence on child nutrition.

11. **What percentage of rural women in India is actively involved in animal husbandry, according to recent surveys?**
a) 2% b) 11%
c) 25% d) 48%

12. **What is the primary advantage of poultry farming for women who stay at home?**
a) Emergency support b) Increased workload
c) Reduced savings d) Decreased income

13. **What does poultry farming help women save money for?**
a) Luxury items b) Family vacations
c) Future investments d) Clothing

14. **What does poultry farming primarily contribute to a household's nutrition?**
a) Vitamins b) Carbohydrates
c) Protein d) Fibre

15. **What distinguishes poultry farming from other livestock activities in terms of women's control?**
a) Women have no control in poultry farming.
b) Women have control over the entire process.
c) Men have control over the entire process.
d) Women only control marketing.

16. **What challenges do women involved in poultry production often face?**
a) Market challenges b) Veterinary challenges
c) Gender equality d) Technological advantages

17. **What is the primary constraint in accessing veterinary services for poultry farming?**
a) Lack of interest b) Lack of access to veterinarians
c) Lack of demand d) Lack of funding

18. **How can women in poultry farming overcome challenges related to knowledge and technology?**
a) By not seeking any knowledge or technology
b) By relying on men for information

c) By attending training programs
d) By avoiding modern technologies

19. **What is one of the key factors that can lead to the success of poultry farming activities?**
a) Lack of training
b) Limited access to credit
c) Market access
d) High mortality rates

20. **What is the primary focus of women's involvement in poultry farming in rural areas?**
a) Meat production
b) Egg production
c) Wool production
d) Milk production

21. **What is the easy source of income for many rural women?**
a) Crop production
b) Off-farm employment
c) Poultry farming
d) Off Season vegetable cultivation

22. **What advantage does poultry farming offer in terms of resource recycling?**
a) It does not involve resource recycling.
b) It efficiently uses household waste.
c) It relies on synthetic feeds.
d) It requires large amounts of water.

23. **How does poultry farming contribute to poverty reduction among rural women and children?**
a) It does not impact poverty reduction.
b) It increases poverty.
c) It has no relation to poverty.
d) It can be a powerful tool for poverty reduction.

24. **What impact does increasing women's income from poultry farming have on their social status?**
a) It reduces their social status.
b) It has no impact on their social status.
c) It often enhances their social status.
d) It lowers their decision-making power.

25. **What are some challenges women involved in poultry production may face related to market access?**
    a) Distance from markets, access to market information, and transport facilities.
    b) Lack of interest in markets.
    c) Overcrowded markets.
    d) Excessive market competition.

26. **What is the key factor that contributes to the ease of managing poultry compared to other livestock?**
    a) Low market demand
    b) High external inputs
    c) Good market prices
    d) Few external inputs

27. **How can training and access to veterinary services improve poultry farming?**
    a) They have no impact on poultry farming.
    b) They can lead to overconsumption.
    c) They can enhance the succcss of poultry activities.
    d) They reduce the need for poultry health services.

28. **What is an essential factor for the success of poultry development in rural areas?**
    a) Lack of credit
    b) Limited access to training
    c) Access to credit
    d) Excessive funding

29. **How does an integrating gender consideration into poultry projects help women?**
    a) It reduces their decision-making power.
    b) It limits their access to resources.
    c) It increases poverty among women.
    d) It identifies factors of production and access to benefits.

30. **What type of policies should be advocated for in support of women in poultry production?**
    a) Policies that do not recognize women's contributions
    b) Policies that overlook the importance of poultry farming
    c) Policies that recognize and support women's crucial role
    d) Policies that promote men's dominance in poultry farming

31. **In poultry farming, what role do women traditionally play in decision-making?**
    a) Women have no role in decision-making.
    b) Women are solely responsible for decision-making.
    c) Women have equal decision-making power with men.
    d) Women often have control over the entire process.

32. **What is the primary advantage of poultry farming in terms of household nutrition?**
    a) It provides essential vitamins only.
    b) It is a source of carbohydrates.
    c) It offers protein-rich food.
    d) It supplies fiber-rich food.

33. **How does poultry farming contribute to resource recycling?**
    a) By using synthetic feeds
    b) By producing excessive waste
    c) By efficiently using household waste
    d) By wasting resources

34. **What does poultry farming often help rural women save money for**?
    a) Luxury items
    b) Family vacations
    c) Future investments
    d) Clothing

35. **What challenges do women involved in poultry production often face related to market access?**
    a) Limited access to vaccines
    b) Lack of interest in markets
    c) Overcrowded markets
    d) Distance from markets and transport facilities

36. **What is a crucial factor for the success of poultry development in rural areas?**
    a) Lack of training
    b) Access to credit
    c) Excessive funding
    d) Lack of interest

37. **On average, how many hours per day do women dedicate to livestock production activities?**
    a. 1-2 hours
    b. 3-5 hours
    c. 6-8 hours
    d. 9-10 hours

38. **In which sector does employ the majority of its workforce as women?**

a. Rice farming
b. Crop production
c. Aquaculture
d. Dairy

39. **In which Indian states is the share of women employment in the livestock sector as high as 90%?**

a. Tamil Nadu and Kerala
b. Punjab and Haryana
c. Gujarat and Rajasthan
d. Karnataka & Andhra Pradesh

40. **What is the estimated number of women engaged in dairying in India?**

a. 1 million
b. 5 million
c. 15 million
d. 75 million

41. **What are some of the feeding-related activities undertaken by women in livestock management?**

a. Slaughtering animals
b. Managing finances
c. Collecting fodder
d. Building shelters

42. **In which aspects of livestock production do women have limited roles?**

a. Taking loans for livestock production
b. Processing dairy products
c. Choosing animals for dairy purposes
d. All of the above

43. **What does livestock ownership offer to women?**

a. Enhanced decision-making power
b. Improved household welfare
c. Increased income generation
d. All of the above

44. **How does livestock ownership contribute to food security?**

a. It ensures a consistent supply of nutritious animal-based foods.
b. It provides access to fresh vegetables.
c. It reduces the need for water resources.
d. It eliminates the risk of crop failure.

45. **What is one of the advantages of poultry farming for women who stay at home?**

a. It offers a means to help the family during times of need.
b. It provides opportunities for international travel.
c. It allows for flexible working hours.
d. It requires minimal effort.

46. **What is a common constraint faced by women in poultry farming?**
    a. Excessive access to veterinary services
    b. Limited access to credit
    c. High market demand for poultry products
    d. Abundance of training opportunities

47. **What impact does poultry farming have on women's social status?**
    a. It has no impact on social status.
    b. It decreases social status.
    c. It enhances social status.
    d. It depends on the location.

48. **What is a primary advantage of poultry farming over other livestock production systems?**
    a. Lower market demand
    b. Lower investment requirement
    c. Greater control by women
    d. Higher resource inputs

49. **What potential shift in roles and attitudes has poultry farming seen?**
    a. A shift towards men taking over poultry farming completely.
    b. A shift towards women gaining more control over income.
    c. A shift towards greater automation in poultry farming.
    d. A shift towards mixed-gender involvement in poultry farming.

50. **What strategy can help address the challenge of limited access to vaccines and veterinary services in poultry farming?**
    a. Providing more access to credit
    b. Establishing better marketing systems
    c. Offering training in husbandry practices
    d. Improving healthcare infrastructure

51. **Which of the following is NOT a recommended measure to support women in poultry production?**
    a. Marketing
    b. Training
    c. Access to credit
    d. Reducing women's control over income

52. **What should policies related to poultry farming aim to do?**
    a. Recognize and support women's crucial role
    b. Exclude women from poultry farming
    c. Promote men's dominance in poultry farming
    d. Limit access to credit for women

53. **What percentage of India's population is employed in the livestock sector?**
    a. 2% b. 8.80%
    c. 15% d. 25.60%

54. **What percentage of the GDP in India is contributed by the livestock sector?**
    a. 1% b. 4.11%
    c. 10% d. 25%

55. **What percentage of all members of dairy producer cooperatives in India are women?**
    a. 5% b. 15%
    c. 25% d. 31%

56. **Which activity involves women spending approximately 1,779 hours per year, significantly more than men?**
    a. Animal Husbandry b. Crop Cultivation
    c. Irrigation d. Duck rearing

57. **What does livestock ownership offer women in terms of decision-making power?**
    a. No significant change
    b. Limited influence
    c. Enhanced decision-making power
    d. Decreased decision-making power

58. **How does livestock ownership contribute to food security?**
    a. It doesn't affect food security.
    b. It reduces food security.
    c. It ensures a consistent supply of nutritious animal-based foods.
    d. It leads to food scarcity.

59. **What is one of the benefits of livestock management in terms of household welfare?**
    a. It doesn't impact household welfare.
    b. It leads to household conflicts.
    c. It contributes to improved overall well-being through income generation and food security.
    d. It decreases overall well-being.

60. **What is a significant source of cash income for women in livestock production?**
    a. Crop farming
    b. Livestock ownership
    c. Daily labor
    d. Government subsidies

61. **What does active involvement in livestock-related activities do for women's self-esteem?**
    a. It doesn't impact self-esteem
    b. It decreases self-esteem
    c. It has no effect on self-esteem
    d. It bolsters self-esteem

62. **Which type of asset is more accessible for women compared to land or financial assets?**
    a. Land
    b. Financial assets
    c. Livestock
    d. Intellectual property

63. **What is essential for effective income management within households?**
    a. Increased spending
    b. Decreased income
    c. Women's control over income
    d. Men's control over income

64. **What is one of the obstacles faced by women in livestock production related to access to resources?**
    a. Overwhelming access to credit
    b. Limited access to credit
    c. Equal access to technology
    d. No access to extension services

65. **What traditional gender roles often confine women to specific tasks within livestock production?**
    a. Leading farm management
    b. Milking cows
    c. Managing finances
    d. Decision-making

66. **What does intensification of livestock production do in terms of women's participation?**
    a. Increases women's participation
    b. Decreases women's participation
    c. Has no impact on women's participation
    d. Changes the type of animals women raise

67. **What is a common issue regarding extension services in the context of women in livestock production?**
    a. Services are highly accessible to women.
    b. Women receive more attention and training.
    c. Extension personnel may not be equipped to engage effectively with women.
    d. Men are excluded from extension services.

68. **What can help challenge and change gender biases and cultural norms in livestock production?**
    a. Ignoring the issue
    b. Providing financial incentives
    c. Awareness campaigns, community dialogues, and targeted interventions
    d. Isolating women from livestock activities

69. **What is the primary workforce involved in the animal husbandry sector in rural areas across agro-ecological zones?**
    a. Men
    b. Children
    c. Both men and women
    d. Women

70. **In the dairy sector, what percentage of the workforce consists of women?**
    a. 50%
    b. 60%
    c. 70%
    d. 80%

71. **What is the main factor that determines the amount of time women spend on livestock-related tasks?**
    a. Household income
    b. Type of animals owned
    c. Educational level
    d. Seasonal availability of fodder

72. **Which of the following activities related to livestock management is primarily carried out by women?**
    a. Buying and selling animals
    b. Taking care of sick animals
    c. Grazing cattle
    d. Milking animals

73. **What is the role of women in processing activities related to animal husbandry?**
   a) Selling milk
   b) Grazing animals
   c) Taking care of sick animals
   d) Veterinary care

74. **What type of tasks are rural women more likely to engage in compared to men in farming activities?**
   a) Administrative tasks
   b) Marketing tasks
   c) More tasks overall
   d) Outdoor activities

75. **What is the trend in the number of women's dairy cooperative societies in India?**
   a) Decreasing over the years
   b) Stable
   c) No data available
   d) Increasing over the years

76. **Which age group of women is more likely to be engaged in livestock rearing?**
   a) Young and educated women
   b) Middle-aged and educated women
   c) Older and less educated women
   d) Women of all age groups equally

77. **In hilly regions in which season do women spend more time on grazing livestock due to the scarcity of fodder?**
   a) Summer
   b) Winter
   c) Spring
   d) Monsoon

78. **Which livestock management activities are predominantly performed by women, except for grazing?**
   a) Taking care of sick animals
   b) Cleaning animal sheds
   c) Selling animals in the market
   d) Providing veterinary care

79. **How does poultry farming contribute to the lives of women in rural areas?**
   a) It decreases their workload by requiring fewer tasks.
   b) It provides them with an opportunity to stay at home.
   c) It reduces their decision-making power in the household.
   d) It enables them to generate income, save money, and improve nutrition.

80. **What is a constraint faced by women in poultry production?**
    a) Lack of interest in poultry farming among women
    b) Availability of excess vaccines and veterinary services
    c) Limited access to market information and transport facilities
    d) Overemphasis on women's control over income

81. **What is a recommended for improving poultry production for women?**
    a) Encouraging men to take over poultry activities completely
    b) Keeping women away from training in husbandry practices
    c) Establishing a well-organized marketing system accessible to women
    d) Ignoring gender considerations in poultry projects

82. **What is one advantage of livestock ownership for women in developing countries compared to other assets?**
    a) Livestock are easier to purchase than land
    b) Livestock provide more immediate financial benefits
    c) Livestock ownership helps women acquire higher education
    d) Livestock ownership increases women's political power

83. **How does livestock ownership by women contribute to household decision-making and spending on children's education and health?**
    a) Livestock ownership reduces women's workload
    b) Livestock ownership decreases women's bargaining power
    c) Livestock ownership increases women's involvement in household decisions.
    d) Livestock ownership has no impact on household dynamics.

84. **Which types of animal products are considered rich sources of energy, proteins, vitamins, and minerals?**
    a) Grains
    b) Vegetables
    c) Animal source foods
    d) Fruits

85. **Why is income management by women crucial for many households?**
    a) Women are more skilled in financial management.
    b) Women spend less money on their families compared to men.
    c) Women's income management can reduce domestic violence.
    d) Men are better at managing income.

86. **Which of the following tasks is NOT associated with manual work in livestock production?**

a) Feeding animals
b) Milking animals
c) Cleaning sheds
d) Using machinery for herding

87. **What is the primary factor that determines women's access to credit in livestock production?**

a) Ownership of large livestock
b) Education level
c) Collateral for loans
d) Gender roles

88. **Why is women's involvement in livestock extension and training limited?**

a) Women are not interested in livestock activities.
b) Extension services are mainly designed for men.
c) Women lack the necessary skills for livestock management.
d) Extension personnel refuse to train women.

89. **What is the significance of income from livestock activities for women's decision-making?**

a) Women tend to use the income for personal expenses.
b) Income from livestock activities is often wasted.
c) Women tend to spend the income on their families' welfare.
d) Women do not have control over income from livestock.

90. **What does the division of labor in livestock production between women and men depend on?**

a) Women's willingness to participate
b) Economic development of the household
c) Ownership of larger animals
d) Women's level of education

91. **What is a common constraint that limits women's participation in livestock activities and training?**

a) Lack of interest in livestock production
b) Men's dominance in decision-making
c) Excessive workload for women
d) Limited mobility due to social and cultural factors

92. **Which aspect of women's safety should be considered in livestock development programs?**
    a) Equitable access to education
    b) Participation in decision-making
    c) Access to credit facilities
    d) Risk of violent assault while fetching water

93. **Involvement of women in which of the following areas is important due to their responsibilities for small and/or young livestock?**
    a) Market research
    b) Crop cultivation
    c) Animal health interventions
    d) Water management

94. **How should veterinary services be designed considering local social and cultural norms?**
    a) Ignoring cultural norms for equality
    b) Conducting training in urban areas
    c) Using male veterinary paraprofessionals only
    d) Tailoring services to accommodate women's limited mobility

95. **What is critical for ensuring women's access to extension services, credit, and technologies?**
    a) Access to formal education
    b) Control over household decisions
    c) Collateral for loans
    d) Ownership of land

96. **How can women become more market-oriented and increase their economic independence?**
    a) By avoiding market-oriented practices
    b) By reducing their involvement in farmers' organizations
    c) By adopting market-oriented practices
    d) By focusing solely on traditional livestock practices

97. **How do social networks contribute to women's livestock farming knowledge?**
    a) By limiting their exposure to new ideas
    b) By encouraging isolation
    c) By promoting modern farming techniques
    d) By limiting their access to information

98. **Why is increasing women's negotiating power and decision-making role important in farmers' organizations?**
    a) To isolate women from decision-making
    b) To discourage their involvement in agricultural practices
    c) To ensure men's dominance in decision-making
    d) To empower women in agricultural matters

Question 2. True or False

1. The amount of time women spend on animal rearing is solely determined by the number and type of animals they own.
2. Women tend to spend more time on animal rearing during seasons when fodder is easily available.
3. Livestock rearing is typically pursued by younger, well-educated women in the village.
4. Women in hilly areas walk significant distances to collect and transport fodder.
5. Women's knowledge of local feed resources enables them to identify suitable plants for animal feed.
6. Young girls are not involved in the grazing of small ruminants in hilly areas.
7. Outdoor activities related to livestock management are primarily controlled by women.
8. Rural women are more involved in marketing activities than processing activities related to livestock products.
9. Women are predominantly responsible for activities like cleaning animal sheds and milking animals.
10. Livestock rearing is primarily a male-dominated activity in rural households.
11. Women's participation in the dairy sector workforce is negligible.
12. About 70 percent of the workforce in the dairy sector is women.
13. The majority of women workers in rural areas are engaged in non-agricultural activities.
14. Women's involvement in the livestock economy is limited to participation in dairy co-operatives.
15. The number of women's dairy cooperative societies in India has decreased over the years.

16. Livestock rearing is found to be an occupation of younger, well educated women.
17. Women in rural areas are mainly responsible for outdoor livestock management activities.
18. Rural women are primarily engaged in marketing activities within the livestock sector.
19. Men are responsible for all livestock management activities except for grazing.
20. Women's involvement in livestock care is limited to providing care for healthy animals.
21. Identifying women's roles as livestock owners is not important for the success of livestock development programs. .
22. Women's safety is not a concern when considering gender roles in livestock programs, as it doesn't affect their participation.
23. Men are rarely involved in animal health interventions and training, as those tasks are traditionally female-dominated.
24. Social and cultural norms have no impact on women's role as service providers in veterinary services.
25. Women's limited access to information and organization doesn't hinder their participation in livestock programs.
26. Women often lack access to capital, knowledge, and technologies in this context.
27. Women's access to assets like water, land, and knowledge is not relevant to their role in decision-making within livestock programs.
28. Women's participation in livestock markets is not important for increasing their income and economic independence.
29. Social networks play no role in improving women's knowledge about livestock farming practices.
30. Farmers' organizations do not need to consider the role of women and their decision-making power.
31. Women's heavy workload doesn't impact their ability to engage in income-generating activities related to livestock.
32. Women's involvement in animal husbandry is not significant and does not contribute much to rural communities and the economy.
33. Livestock production in India is primarily dominated by men.
34. Women's contributions in livestock care are limited to feeding and general management tasks.

35. Women who engage in livestock production often have limited access to resources such as credit and technology.
36. In some regions, women make up the majority of the labor force in livestock farming.
37. Livestock ownership by women can lead to improved decision-making within households.
38. Empowering women through livestock ownership has no impact on food security.
39. Women's involvement in livestock production can lead to increased income generation.
40. Women are generally not involved in processing and marketing activities related to livestock products.
41. The role of women in livestock management is consistent across all regions and communities.

Question 3. Fill in the blanks

1. Livestock ownership is advantageous for women in developing countries as they can more easily own __________.
2. Animal source foods, such as milk, meat, and eggs, are rich in energy and provide a good source of __________.
3. Women's ability to manage their income is crucial for many households, as studies show they spend close to __________ of their income on their family.
4. Livestock ownership by women increases the probability that households, especially children, will benefit from the consumption of livestock products or from food bought using income derived from __________.
5. Women's ownership of livestock increases their role in household decision-making and household spending on __________.
6. Studies have shown that women are more likely to own and benefit more from __________ rather than larger animals.
7. Livestock ownership can help women move out of __________.
8. Interventions that increase women's access and rights to livestock, and then safeguard them from dispossession and theft, contribute to helping women improve their __________.
9. Women's management of income from livestock sales can reduce domestic violence and improve the __________ of their children.

10. Women play a pivotal role in ___________, particularly in the context of India, contributing significantly to rural communities' well-being and the nation's economy.
11. Despite their crucial roles, women face numerous obstacles and constraints in livestock production, including limited access to resources, gender biases, and ___________.
12. Livestock ownership offers benefits to women, including enhanced decision-making power, improved household welfare, increased income generation, and heightened ___________.
13. Women often face challenges in poultry farming, such as limited access to markets, healthcare, knowledge, technology, and ___________.
14. To support women in poultry production, it is essential to establish well-organized ___________ accessible to women to ensure better prices for poultry products.
15. Providing training in husbandry practices and improving access to poultry health services can enhance the success of poultry activities and overcome ___________.
16. Availability of ___________ is often essential for poultry development in rural areas, enabling women to invest in quality feed and medicines.
17. Incorporating gender considerations into poultry projects helps identify factors of production and access to benefits, promoting ___________.
18. Advocating for policies that recognize and support the crucial role women play in poultry production and rural development involves policymakers and planners in ___________.
19. Poultry farming is unique because, unlike other livestock production systems, rural women often have control over the entire process, from feeding the birds to ___________.

ANSWERS

1. Multiple Choice Questions

1	d	1,779 hours
2	d	55.3%
3	d	Dairy farming
4	a	Enhanced decision-making power
5	b	By providing a consistent supply of nutritious animal-based foods

6	b	It contributes to overall well-being.
7	c	It serves as a reliable source of cash income.
8	d	It bolsters self-esteem
9	b	More accessible
10	c	It often leads to improved resource allocation.
11	b	11%
12	a	Emergency support
13	c	Future investments
14	c	Protein
15	b	Women have control over the entire process.
16	a	Market challenges
17	b	Lack of access to veterinarians
18	c	By attending training programs
19	c	Market access
20	b	Egg production
21	c	Poultry farming
22	b	It efficiently uses household waste
23	d	It can be a powerful tool for poverty reduction
24	c	It often enhances their social status
25	a	Distance from markets, access to market information, and transport facilities
26	d	Few external inputs
27	c	They can enhance the success of poultry activities
28	c	Access to credit
29	d	It identifies factors of production and access to benefits
30	c	Policies that recognize and support women's crucial role
31	d	Women often have control over the entire process.
32	c	It offers protein-rich food
33	c	By efficiently using household waste
34	c	Future investments
35	d	Distance from markets and transport facilities
36	b	Access to credit
37	b	3-5 hours
38	d	Dairy

39	b	Punjab and Haryana
40	d	75 million
41	c	Collecting fodder
42	d	All of the above
43	d	All of the above
44	a	It ensures a consistent supply of nutritious animal-based foods
45	a	It offers a means to help the family during times of need
46	b	Limited access to credit
47	c	It enhances social status
48	c	Greater control by women
49	d	A shift towards mixed-gender involvement in poultry farming
50	d	Improving healthcare infrastructure
51	d	Reducing women's control over income
52	a	Recognize and support women's crucial role
53	b	8.80%
54	b	4.11%
55	d	31%
56	d	Feeding animals
57	c	Enhanced decision-making power
58	c	It ensures a consistent supply of nutritious animal-based foods
59	c	It contributes to improved overall well-being through income generation and food security
60	b	Livestock ownership
61	d	It bolsters self-esteem
62	c	Livestock
63	c	Women's control over income
64	b	Limited access to credit
65	b	Milking cows
66	b	Decreases women's participation
67	c	Extension personnel may not be equipped to engage effectively with women
68	c	Awareness campaigns, community dialogues, and targeted interventions
69	d	Women
70	c	70%

71	b	Type of animals owned
72	d	Milking animals
73	a	Selling milk
74	c	More tasks overall
75	d	Increasing over the years
76	c	Older and less educated women
77	b	Winter
78	b	Cleaning animal sheds
79	d	It enables them to generate income, save money, and improve nutrition
80	c	Limited access to market information and transport facilities
81	c	Establishing a well-organized marketing system accessible to women
82	a	Livestock are easier to purchase than land
83	c	Livestock ownership increases women's involvement in household decisions
84	c	Animal source foods
85	c	Women's income management can reduce domestic violence
86	d	Using machinery for herding
86	c	Collateral for loans
87	b	Extension services are mainly designed for men
88	c	Women tend to spend the income on their families' welfare
89	b	Economic development of the household
90	d	Limited mobility due to social and cultural factors
91	d	Risk of violent assault while fetching water
92	c	Animal health interventions
93	d	Tailoring services to accommodate women's limited mobility
94	c	Collateral for loans
95	c	By adopting market-oriented practices
97	c	By promoting modern farming techniques
98	d	To empower women in agricultural matters

Answer 2. True and False

S.No.	True or false	S.No.	True or false
1	False	2	False
3	False	4	True
5	True	6	False
7	False	8	False
9	True	10	False
11	False	12	True
13	False	14	False
15	False	16	False
17	False	18	False
19	False	20	False
21	False	22	False
23	False	24	False
25	False	26	True
27	False	28	False
29	False	30	False
31	False	32	False
33	False	34	False
35	True	36	True
37	True	38	False
39	True	40	True
41	False		

Answer 3. Fill in the Blanks

1	Livestock assets
2	Proteins, vitamins, and minerals
3	90%
4	Livestock
5	Children's education and health
6	Small stock
7	Poverty
8	Bargaining power
9	Nutritional status

10	Animal husbandry
11	Cultural norms
12	Self-esteem
13	Credit
14	Marketing systems
15	Healthcare challenges
16	Credit
17	Gender integration
18	Policy support
19	Marketing the products

3

Importance of Gender Sensitization in Animal Husbandry

Chapter Overview

This chapter discusses the crucial concept of gender sensitization in the context of livestock production. Gender sensitization involves raising awareness about gender equality concerns, challenging stereotypes, and promoting equitable practices in the livestock sector. The chapter covers the stages of gender sensitization, offers guidelines for veterinarians and veterinary students, provides real-world examples, outlines a sensitization strategy, and underscores the pressing need for gender sensitization in livestock production. It highlights how this process can lead to increased productivity, sustainability, and well-being in the industry by breaking down traditional gender roles and fostering inclusivity and cooperation among all stakeholders.

SECTION A: THEORY

Definitions of Gender Sensitisation

Gender sensitizing "is about changing behaviour and instilling empathy into the views that we hold about our own and the other sex." It helps people in "examining their personal attitudes and beliefs and questioning the 'realities' they thought they know.

(Aksornkool *et al.*, 2008)

Gender sensitisation is the modification of behaviour through awareness of gender equality concerns. It is the teaching of gender sensitivity and encouragement of behaviour modification through raising awareness of gender equality concerns.

(Sharma, 2014)

Gender Sensitization for Veterinarians

i. **Awareness:** The first step in gender sensitization is to make veterinarians aware of the existing bias and stereotypes related to gender in the field. They need to recognize that these biases might be affecting their career opportunities and professional relationships.

ii. **Empathy and Understanding:** Veterinarians should be encouraged to put themselves in the shoes of their colleagues of the opposite gender. This means understanding the challenges and biases that their colleagues might face, such as the assumption that women are not as capable of handling large animals as men.

iii. **Behavior Modification:** To promote gender equality and fairness, veterinarians should actively work towards behavior modification. For example, male veterinarians can actively seek opportunities to work with smaller pets and highlight their expertise, challenging the stereotype that they are only suited for large animals. Female veterinarians can similarly take on more significant animal cases to demonstrate their abilities.

v. **Education and Training:** Veterinarians should receive training and participate in workshops on gender sensitization. They can learn about the importance of diversity and inclusivity, how biases can affect workplace dynamics, and strategies to promote equality.

Supreme Court of India released a Handbook on Gender Stereotypes in recognizing, understanding, and combating gender stereotypes present in legal language and judgments. The handbook also covers common stereotypes about gender roles incorrect notions about women's roles and behavior. Some of these are discussed below:

Common stereotypes about the gender roles ascribed to men and women, and why they are incorrect

Stereotype	**Reality**
1.Women are more nurturing and better suited to care for others.	People of all genders are equally suited to the task of caring for others. Women are often socially conditioned to care for others from a young age. Many women are also forced to abandon their careers to care for children and the elderly.
2.Women should do all the household chores.	People of all genders are equally capable of doing house chores. Men are often conditioned to believe that only women do household chores.
3.Wives should take care of their husband's parents.	The responsibility of taking care of elderly individuals in the family falls equally on individuals of all genders. This is not the sole remit of women.
4.Women who work outside of the home do not care about their children.	Working outside of the home has no correlation with a woman's love or concern for her children. Parents of all genders may work outside of the home while also caring for their children.

5.Women who are also mothers are less competent in the office because they are distracted by childcare.	Women who have "double duty," working outside the home and raising children, are not less competent in the workplace.
6.Women who do not work outside the home do not contribute to the household or contribute very little in comparison to their husbands.	Women who are homemakers perform unpaid domestic and care work, contributing to the household's quality of life and resulting in monetary savings. Their contributions are often overlooked due to societal conditioning.

Stereotypes based on the so-called "inherent characteristics" of women

Stereotype	Reality
Women are overly emotional, illogical, and cannot take decisions.	A person's gender does not determine or influence their capacity for rational thought.
All women are physically weaker than all men.	While men and women are physiologically different, it is not true that all women are physically weaker than all men. Strength depends on various factors including profession, genetics, nutrition, and physical activity, not solely on gender.
Women are more passive.	People display a wide range of personality traits. Both men and women can be (or may not be) passive. Women are not more passive than men as a rule.
Women are warm, kind, and compassionate.	Compassion is an acquired characteristic unique to every individual. Individuals of all genders can possess (or not possess) compassion.
Unmarried women (or young women) are incapable of taking important decisions about their lives.	Marriage has no bearing on an individual's ability to make decisions. Legal consent ages for activities like marriage or alcohol consumption apply irrespective of marital status.
Women of oppressed or marginalised communities have diminished cognitive capabilities or a limited understanding of the world.	The community an individual belongs to does not determine their cognitive capabilities or understanding of the world.
All women want to have children.	Not all women want to have children. Parenthood is an individual choice based on various circumstances.

Stages of Gender Sensitisation

Gender sensitization process involves four stages; change in perception, recognition, accommodation, and action. These changes take place in response to certain interventions i.e., sensitization training.

i) Change in perception

Gender sensitization initiates us to think about gender differently. In first instance, it tends to change the perception that men and women have of each other. It creates a mind-set in men that no longer sees in women the stereotypical image. The impression that women are a weak and unequal identity no more clouds the minds of commoners. Rather, women are seen as responsible and equal partners in socio-economic development.

Guidelines on gender sensitive approach for veterinary students

i. **Equal treatment**: Veterinary students need to treat male and female livestock farmers and pet owners equally. This means giving equal weight to their concerns, opinions, and knowledge about their animals. For example, if a female livestock farmer expresses concern about her herd's health, the student should take it as seriously as they would if a male farmer raised the same concern.

ii. **Encouraging female participation**: Gender sensitization also involves encouraging female livestock farmers and pet owners to actively participate in discussions and decision-making regarding their animals. Students can create an inclusive environment where everyone's voice is heard. For instance, during a community animal health workshop, female pet owners should be encouraged and provide opportunity to share their experiences and insights just as much as male owners.

iii. **Education and training**: Veterinary students can be trained to provide information and education in a gender-sensitive manner. For example, when discussing livestock management practices, students can tailor their advice to the specific needs and circumstances of female livestock farmers. They can consider factors like women's schedules, responsibilities, and available resources while designing a training programme for women. It is worth gender sensitization training for both men and women to address gender issues.

iv. **Capacity building of women**: Veterinary students can also empower female livestock farmers and pet owners by helping them acquire essential knowledge and skills. This can include training women in basic animal healthcare, nutrition, or disease prevention. Empowering women in these ways can lead to improved animal care and overall community development.

v. **Challenging stereotypes**: Gender sensitization can involve challenging and breaking down stereotypes in the field. For example, students should be encouraged to see female livestock farmers as equally capable of handling tough tasks, such as assisting with animal birthing or administering medications.

vi. **Flexible services**: Veterinary services can be made more accessible to both male and female livestock farmers and pet owners. This could involve offering evening or weekend clinics to accommodate various schedules, ensuring that women have equal access to veterinary care.

vii. **Data collection and research**: Students can be encouraged to collect gender-disaggregated data related to animal health and management. This data can help identify specific challenges or differences in the ways men and women care for their animals and can inform more targeted support and interventions.

By incorporating these practices into their work, veterinary students can contribute to a more gender-sensitive and equitable approach to animal care, benefiting both male and female livestock farmers and pet owners while improving the well-being of animals in their care.

ii) Recognition

i. Persons exposed to gender sensitization try to look at the positively endowed qualities of women.

ii. At this stage, the male folk come around to recognize the virtues of women and their importance to the family and the society.

iii. There is spontaneous appreciation for women's involvement in multifarious activities.

iv. As a result, women's contributions become increasingly visible. Further, women's talents and capabilities that were going unnoticed and unexplored become subject of attention.

v. Given an opportunity, women too become more conscious of their capability and contribution and take pride in the same.

vi. Women, cutting across socio-economic boundaries, tend to see their problems in larger perspective of women development and come forward to recognize the efforts of fellow women.

vii. They even visualize the key role that men can play in their socio-economic development. Men need to recognise and appreciate the contribution of women to their family welfare.

Points to ponder for Veterinary Students

i. **Acknowledging Women's Integral Role:** Veterinary students, after undergoing gender sensitization, become more aware of the significant role women play in livestock farming. They recognize that women often manage the daily care of animals, including feeding, milking, and tending to their healthcare needs. For instance, they appreciate that women in rural areas are key caregivers to dairy cows, ensuring the animals' health and productivity. They must understand that both men and women in a farming family need to share their responsibilities for wellbeing of their families.

ii. **Recognizing Women as Pet Owners:** Through gender sensitization, veterinary students also come to acknowledge that women are often primary caregivers for pets in households. They recognize the role women play in ensuring the well-being of pets, which includes feeding, grooming, and taking them to the veterinarian. For example, they understand that women may schedule and attend vet appointments for family pets.

iii. **Appreciating Women's Expertise:** As part of the recognition phase, veterinary students understand that women, especially in rural areas, possess valuable knowledge about animal husbandry and pet care. For instance, they might learn from women in farming communities about traditional practices for animal healthcare, which have been passed down through generations. By virtue of their close association with livestock, women are the first to identify sick animals and animal in heat and assess the quality of feed they are giving to the animals.

iv. **Promoting Women's Involvement:** Gender sensitization encourages veterinary students to actively involve women in discussions and decision-making related to livestock farming and pet ownership. They seek out women's input when providing advice on animal health and management. This could involve consulting with women farmers on the best practices for disease prevention and treatment. To promote clean milk production practices (CMP) it is better to train women on CMP as they are involved in shed hygiene, cleaning utensils, milking of animals etc.

v. **Celebrating Women's Achievements:** Veterinary students may take the initiative to highlight and celebrate the accomplishments of women in animal care and agriculture. They may organize events or campaigns to recognize and appreciate women's contributions, such as featuring successful female livestock farmers or pet owners in local media.

vi. **Empowering Women in Veterinary Sciences:** Gender sensitization can inspire veterinary students to support and mentor women who aspire to become veterinarians or pursue careers in animal care. They can provide guidance and encouragement to women interested in the field, helping them overcome barriers and biases. It is a good sign that today at least 50% of the students who are pursuing veterinary education are women.

vii. **Fostering Collaboration:** With a more gender-sensitive perspective, veterinary students can collaborate with women's groups and organizations that focus on livestock farming and pet care. This collaboration can lead to improved knowledge exchange and joint initiatives aimed at enhancing animal health and welfare.

By recognizing and appreciating the roles of women in livestock farming and pet ownership, veterinary students can contribute to a more inclusive and effective approach to animal care and agriculture that benefits both the animals and the communities they serve.

iii) Accommodation

i. The barrier between men and women starts crumbling down in real sense and the society slowly gets over the perennial problem of adjustment between them. For example, in the veterinary field, there has been a traditional perception that it's more common for men to deal with livestock marketing so most times decisions regarding sale and purchase may be undertaken by men. However, this barrier needs to break down as at times women may benefit more when they are involved in this domain.

ii. Men tend to rationalize their behaviour by burying their ego as far as gender relations are concerned. For example, a male veterinarian may be open to seeking advice or collaborating with a female colleague in cases where the latter has expertise or experience in a specific area, like animal behaviour. He doesn't let his ego hinder his ability to provide the best care for the animals.

iii. Instead of complaining or reacting to the behaviour of women, men learn to exercise patience and restraint, and take the things in a positive way. For example, a male veterinarian may find himself working with a female livestock farmer who has different ideas about animal husbandry. Instead of reacting negatively to these differences, he chooses to be patient and communicate effectively with the women farmers to find common ground for the well-being of the animals.

iv. In the family, women start gaining importance as their opinions and suggestions are counted for overall development and management of family. For instance, the wife may recommend changes in the animal diet or housing conditions to improve the health and productivity of the livestock. Her suggestions are valued, and the family collectively decides to implement these changes, leading to better overall outcomes for the animals and the family's livelihood.

v. At community and organizational level too, women are encouraged to play their role in matters of management. Women, on their part, tend to underplay the problems with their male counterpart and wish to solve their problems through dialogue. In a veterinary organization, women are actively encouraged to take leadership roles in decision-making and management positions. When conflicts arise, women and men work together through open and constructive dialogue to find solutions, fostering a harmonious and productive working environment.

Points to ponder for Veterinary Students

i. **Breaking Gender Barriers:** In the realm of veterinary medicine, male and female students can work together to assist livestock farmers in managing their herds. They can collaborate on educational programs and workshops where both genders contribute equally, helping farmers overcome traditional gender biases.

ii. **Rationalizing Behaviour:** Veterinary students can demonstrate gender-neutral behaviour when dealing with livestock farmers and pet owners. They can prioritize the health and well-being of the animals over personal biases, helping to rationalize and normalize gender relations within the field.

iii. **Exercising Patience and Restraint:** Veterinary students should exercise patience and restraint when working with clients, regardless of gender. For example, if a male student encounters a female pet owner who may be overly concerned about her pet's health, he can respond with empathy, addressing her concerns and questions calmly.

iv. **Empowering Women's Voices:** In the context of family-owned farms or pet care, veterinary students can encourage female family members to actively participate in animal care decisions. They can provide information and support to ensure that everyone's opinions are considered, thus promoting the overall development and management of the animals.

v. **Promoting Dialogue:** Veterinary students can facilitate open communication between male and female livestock farmers or pet owners. They can act as mediators to resolve disputes or concerns, fostering a more cooperative and solution-oriented approach to managing animals and their health.

These examples demonstrate how veterinary students can apply the principles of accommodation to their interactions with livestock farmers and pet owners. By promoting gender equality, open communication, and empathy, they can contribute to the well-being of animals and foster positive relationships within the community.

iv) Action

i. *Gender sensitized persons as instruments of change:* Individuals who have been sensitized to gender issues can play a pivotal role in improving the status of women in society. These individuals are more aware of gender inequalities and actively work to create a more favourable environment for women. For example, Veterinarians who have been sensitized to gender issues can positively impact their work with livestock farmers.

   Two examples:

   Veterinarians Working with Livestock Farmers: A female veterinarian in a rural area where livestock farming is a common occupation realizes that female farmers often face unique challenges in accessing veterinary services due to traditional gender roles. She decides to act and organizes a series of workshops and training sessions specifically for female livestock farmers. These workshops empower women with knowledge and skills to better care for their animals. The female veterinarian also collaborates with local women's groups to provide affordable veterinary services to female farmers. This action not only improves the health and well-being of the livestock but also promotes gender equality by breaking down traditional barriers.

   Veterinarians Working with Pet Owners: Veterinarians working with pet owners can also play a role in promoting gender sensitivity. Here is an example:

   A male veterinarian in a busy urban clinic notice that female pet owners often face challenges in finding time to bring their pets in for routine check-ups due to juggling work, childcare, and household responsibilities. To address this, he decides to extend the clinic's operating hours, offering evening and weekend appointments to accommodate the schedules of working women. Additionally, he provides online consultation services to offer convenience to pet owners. This not only ensures better healthcare for the pets but also supports gender equality by recognizing and addressing the specific needs of female pet owners.

ii. *Affirmative Actions to Improve Women's Conditions:* This point emphasizes the need for policies and programs that promote women's participation in development and decision-making processes. Such actions aim to ensure that women have equal opportunities and can share the benefits of development equitably.

   *Example:*Create financial assistance programs or grants for women interested in starting their veterinary practices. Reducing the financial barriers to entrepreneurship can encourage more women to become independent veterinary service providers.

Establish community veterinary clinics or mobile veterinary services that are easily accessible to women, especially in rural areas. These services should be affordable and designed to meet the specific needs of female farmers and pet owners.

ii. Actions could also be in the form of research and extension initiatives to reach out to the women with appropriate technologies and institutional innovations.

Researchers could conduct studies to identify and develop better livestock breeding techniques that are more suitable for small-scale women farmers. This research may lead to the creation of new livestock breeds or genetic improvements that enhance the productivity and resilience of animals in female-led households.

Creating women-centric cooperatives or self-help groups where female livestock farmers can collectively access resources, funding, and knowledge. These institutions can offer financial services, share resources like feed or equipment, and provide a platform for women to voice their concerns and ideas. These innovations can enhance the social and economic status of women in the livestock sector.

Gender Sensitization Strategy

Gender sensitization strategy involves three components:

i. Selecting the target audience

ii. Deciding the content

iii. Deciding the methodology

i) Selecting the target audience

i. *Sensitizing Men and Women in Livestock Farming Communities:* In rural livestock farming communities, men traditionally dominate this sector. Sensitization programs should target men to make them aware of the benefits of involving women in livestock management. These programs can highlight how women's participation can lead to increased household income and improved livestock management practices. For instance, workshops and awareness campaigns can be organized to educate men about how women's contributions, such as animal healthcare or dairy product processing, can enhance overall productivity and sustainability in the sector.

At the same time, it's crucial to sensitize women who may initially resist taking on more active roles due to cultural norms. For instance, some women may be hesitant to participate in decision-making related to livestock. They may be encouraged to attend training programs that demonstrate how their involvement can lead to improved animal welfare and household income.

ii. *Educating Elderly Women about Gender Bias:* In many rural areas, elderly women often play a significant role in upholding traditional gender roles and norms. Sensitization programs can educate elderly women about the negative consequences of gender bias in livestock farming. For example, they can be informed about how restricting women's access to resources or decision-making power can hinder the progress of the entire household.

   An example might involve organizing community discussions and storytelling sessions in which elderly women can learn about successful cases of gender-inclusive livestock management and the positive outcomes it has had on their communities. This can help them see the benefits of changing their attitudes towards younger women's participation in the sector.

iii. *Engaging Socially and Economically Progressive Women:* Socially and economically progressive women in a village or locality can serve as role models and mentors for underprivileged women in the livestock sector. Sensitization programs can target these progressive women to encourage them to support and empower the underprivileged women in their community. For instance, they can establish women's cooperatives or self-help groups focused on livestock rearing and provide training and resources to those in need.

iv. *Sensitizing Researchers, Policy Makers, and Service Providers:* Researchers, policy makers, and service providers play a crucial role in shaping the livestock sector through policies and services. Sensitization programs for this group could involve training sessions and seminars. For example, researchers can be educated about the importance of gender-inclusive research in livestock farming, and how this can lead to more effective and sustainable practices. Policy makers can be sensitized to the need for gender-sensitive policies in the sector. Service providers can learn how to offer support that caters to the specific needs of women in livestock management. Deliberately involving women as members of the gender research projects, PRA teams etc go a long way in sensitizing different stakeholders in livestock development.

ii) Content of the programme

i. Content should amply communicate the intended message to the audience and should be easily understandable by them. In the livestock sector, this could mean creating educational materials or workshops that are accessible to both male and female farmers. For example, creating simple, visually appealing guides on best practices for livestock care that are available in local languages and use clear images to convey information.

ii. Contents of the programme can be decided depending on its very purpose. In the livestock sector, the purpose could be to promote gender equality and women's participation in livestock farming. Content might include training sessions on animal husbandry and farm management. These could be tailored to address the specific needs and goals of the community, whether it's improving milk production or sustainable livestock farming.

iii. It could be to sensitize people about ill-effects of gender bias and discriminatory practices on women, men, family, and society. In the livestock sector, this would involve addressing biases that limit women's access to resources and opportunities in farming. For example, workshops or awareness campaigns could highlight how women's involvement in livestock farming can lead to increased household income and nutrition, and how gender bias can hinder these benefits.

iv. Gender sensitization may focus on spreading the message 'how women play important role in family and in the society' and 'how both men and women in their mutually supportive role can contribute immensely to family welfare, growth, and development of their villages. In the livestock sector, this means emphasizing the shared responsibility of men and women in managing livestock. For instance, creating content that highlights the vital roles women play in livestock care, like milking and animal healthcare, and how these contribute to the overall well-being of the family and community.

v. Contents should initiate friendly debate among larger audience on the ill effects of different forms of gender bias and what can be done to remove such biases. It can focus on the conduct of men and women in a household based on case studies and even spread the massage of some kind of affirmative action. In the livestock sector, this could involve organizing community discussions or sharing case studies of households that have successfully promoted gender equality in livestock farming. These discussions might explore the challenges women face in accessing resources and participating in decision-making, and how affirmative actions, such as providing women with access to training and resources, can help address these biases.

iii) Methodology

This requires following:

i. **Gender-sensitive modules containing case studies; situation analysis, etc.:** In the livestock sector, a gender-sensitive module might include case studies that highlight the roles and contributions of both men and women in livestock management. For example, a case study could focus on a family where the woman is responsible for poultry management, while the man takes care of larger livestock like cows. By analyzing such situations, stakeholders can understand how gender dynamics play out in livestock farming.

ii. **Gender-sensitive materials such as leaflets, booklets, posters, and videos:** These materials could be designed to educate people about gender equity in the livestock sector. For instance, a poster could depict men and women working together in various livestock-related tasks to emphasize the importance of gender equality. A video could highlight successful women in the livestock industry who have made significant contributions to livestock farming.

iii. **Organization of sensitization camps:** Sensitization camps could bring together members of the livestock community to discuss gender issues. Participants might engage in activities that promote understanding, like role-playing scenarios that illustrate how unequal distribution of livestock-related tasks can affect both men and women. These camps could encourage open dialogue and problem-solving related to gender biases in the sector.

iv. **Sustained campaign by mass media and plays:** Mass media campaigns in the livestock sector could feature advertisements and programs that challenge stereotypes and highlight successful stories of gender equality. Additionally, theater plays can be organized to present real-life scenarios, such as a play that portrays a family overcoming gender bias by sharing livestock responsibilities and benefiting from it.

v. **Participatory discussions among men and women from different age groups and the same households:** These discussions can involve both men and women of various age groups who are engaged in livestock farming within the same household. For example, in a participatory discussion, a young woman might share her experiences of successfully introducing new livestock management practices, while an older man could discuss how he has adapted his methods to accommodate gender equality. These dialogues foster understanding and provide a platform for sharing practical experiences. In the livestock sector, promoting gender sensitivity is essential not only for achieving gender equity but also for enhancing the productivity and sustainability of livestock farming. These strategies and examples can help create awareness and drive positive changes in the industry.

Gender-sensitive agricultural extension – key shifts required

Initiating gender sensitive agricultural extension would necessitate key shifts in some of the traditional approaches to extension.

	From	**To**
Objective	Increasing production and productivity	Improved income and more productive employment opportunities for both genders
	Forming self-help groups (SHGs)	Forming common-activity groups open to both genders
	Distribution of inputs	Development of local capacity for sustained availability of inputs and services
Selection of Interventions	Selection of interventions based on Participatory Rural Appraisal (PRA)	Demand-led and based on analysis of men and women client data, matched with opportunities and availability of complementary support and services
	Centrally designed ideas	Client aspirations carefully analyzed with local and external knowledge and support
Approaches	Fixed or uniform	Diverse and gender-differentiated
Working with	Women	Working in partnership with all actors who could support rural women
Monitoring & Evaluation	Input and Output targets	Behavioral and livelihood changes in men, women, youth and other clients and related organizations
	Subjective evaluations	Objective evaluations on a scientifically validated benchmark data basis,
Targeting the poor	Inclusion by accident	Inclusion by design

Source: CRISP 2007 adapted by Blum *et al.*

Such key shifts should be the guiding principles for initiating gender-sensitive agricultural extension reforms. However, to do this, extension must embrace a learning-based approach.

Gender-sensitive Livestock extension advisory services adapted from above table.

A. Change in Objectives

i. From merely enhancing production and productivity to facilitating improved income and the creation of more productive employment opportunities for both genders

This transition represents a fundamental shift from a singular emphasis on boosting production to a more comprehensive goal of enhancing income and fostering more productive employment opportunities for individuals of all genders. Recognizing the multifaceted impact of veterinary and animal husbandry practices on the socio-economic landscape, this broader objective acknowledges that success in livestock extension goes beyond mere output figures. Instead, it seeks to ensure sustainable and inclusive growth, emphasizing equitable economic opportunities and improved livelihoods for both men and women involved in the livestock sector. This evolution aligns with the understanding that the impact of livestock practices extends to the socio-economic well-being of communities, making the objective more holistic and gender sensitive.

ii. Change in approach from creating self-help groups (SHGs) to establishing common-activity groups open to individuals of all genders

Encouraging the establishment of common-activity groups, rather than exclusive self-help groups, fosters collaborative engagement. This shift ensures shared responsibilities in veterinary and animal husbandry activities, promoting inclusivity and equitable participation for both men and women.

iii. Distribution of Inputs: from distribution of inputs to development of local capacity for sustained availability of inputs and services

Moving away from a focus solely on input distribution, the emphasis is on developing local capacity. This involves building the skills and knowledge within the community for the sustainable availability and management of inputs and services in veterinary and animal husbandry extension.

B. Selection of Interventions

a. Shift in selection of interventions based on Participatory Rural Appraisal (PRA) to demand-led and based on analysis of men and women client data, matched with opportunities and availability of complementary support and services.

Transitioning from a reliance on Participatory Rural Appraisal (PRA) to a demand-led approach signifies a shift towards interventions that are more responsive to the specific needs of men and women. This change involves analyzing client data disaggregated by gender, aligning interventions with identified opportunities, and ensuring the availability of complementary support and services in veterinary and animal husbandry extension.

b. Change in designing central ideas to client aspirations carefully analyzed with local and external knowledge and support.

This involves shifting from centrally designed ideas to a meticulous analysis of client aspirations involves understanding the unique needs and goals of the community. This approach considers both local and external knowledge, ensuring that veterinary and animal husbandry extension services are tailored to the specific context and desires of the clients. These shifts contribute to a more gender-sensitive approach in agricultural and animal husbandry extension by prioritizing client-driven interventions. By considering the aspirations and needs of men and women separately, and incorporating local knowledge into program design, extension services become more relevant, effective, and responsive to the diverse requirements of the community.

C. Working with Women

I. Shift from fixed approach of working with men/women towards an evolving, diverse and gender-differentiated approach

In a fixed or uniform approach to agricultural extension, the services and advice provided are often generic and not tailored to the diverse needs of diverse groups within the community. This might result in overlooking the specific challenges and opportunities faced by men and women in animal husbandry. Deeply rooted in the old mentality is also the perception that if extension services are given to a member of the family (mostly the male head of household), then the information will trickle down to the rest of the household, including female members. However, this is not always the case, as men do not necessarily discuss production decisions with their wives, or share information and knowledge with them (Ragasa, 2014). Shifting to an evolving, diverse, and gender-differentiated approach involves customizing extension services based on the unique requirements and roles of men and women in animal husbandry. For example, instead of generic training on animal health, extension services could evolve to provide tailored training programs. For instance, women might receive specific training on issues related to small livestock management, reproductive health of animals, and disease prevention strategies that align with their responsibilities.

II.Shift in Working with Women towards Working in Partnership:

Moving from a fixed approach of working with women to a more inclusive partnership involves recognizing women as active contributors to decision-making processes related to animal husbandry. For example, eestablishing women's cooperatives focused on animal health. These cooperatives could actively participate in decision-making related to animal healthcare practices, disease prevention, and collectively accessing veterinary services. Likewise, actively involving women in community-level committees related to animal health. This ensures that women's perspectives are considered in decisions related to disease control programs, vaccination schedules, and overall animal husbandry practices.

D.Monitoring and Evaluation

i. Shift from Input and Output targets to Behavioural and livelihood changes in men, women, youth & other clients and related organizations

Evaluation of veterinary extension programs may be based solely on the number of vaccinations conducted or diseases treated, without considering the broader impact on the livelihoods of men and women involved in animal husbandry. However, a shift is required by adopting a gender-sensitive evaluation approach. This assesses not only the quantitative outputs but also the behavioral and livelihood changes. The evaluation is expanded to include not only the vaccination count but also the subsequent impact on the community. It assesses changes in livestock health, milk production, and income levels. Importantly, it examines how women and youth have been involved in decision-making related to animal health and management, leading to broader changes in household dynamics and economic well-being.

ii. Change in Subjective evaluations to objective evaluations on a scientifically validated benchmark data basis, considering gender and wealth

Assessment of the success of veterinary extension programs may rely solely on expert opinions or anecdotal evidence, without systematically capturing the experiences and perspectives of diverse stakeholders. However, shift is required by introduction of objective evaluations based on scientifically validated benchmark data. This could involve conducting surveys and studies to gather quantifiable information on the impact of extension services. For instance, assess the prevalence of diseases before and after the implementation of a program and correlate these findings with the socio-economic status of diverse groups, including gender and wealth categories.

E. Targeting the Poor

i. Inclusion by Chance to Inclusion by Design

Inclusion of marginalized communities in veterinary extension activities might happen incidentally, without intentional efforts to address their specific challenges to implement programs that intentionally target the poor and marginalized groups. For instance, design extension services that consider the specific needs of small-scale women farmers or landless households relying on livestock for their livelihoods. Provide targeted support, such as affordable veterinary care or access to improved animal husbandry practices, to address the unique challenges faced by these groups.

In summary, these examples highlight the need for a more inclusive, gender-sensitive, and targeted approach in veterinary and animal husbandry extension. By actively involving women, adopting comprehensive evaluation strategies, and intentionally targeting marginalized communities, extension services can better address the diverse needs and realities within the sector.

i. **African Women Leaders in Agriculture and Environment (AWLAE) program:** The following two programmes were initiated in Africa to develop the capacities of women extension advisors.

ii. **Implementer:** Winrock International

iii. **Donors:** USAID and others

iv. **Goal:** Increased participation and leadership of African women at all levels of agricultural and environmental activities, including production, extension, research, agricultural university teaching, and policymaking.

v. **Approach:** Institutional change; human capacity development for women and men; linkages with rural women.

vi. **Strategic interventions:**

- Scholarships to women for advanced degree training and leadership training.
- Documentation of the status and issues of professional women.
- Engagement with policymakers for policy advocacy.
- Gender sensitization and training for women and men to address gender issues.
- Gender reviews of policies, programs, and curricula.
- Building gender-sensitive male role models for gender advocacy.
- Creation of professional women's associations.
- Linking professional women to women farmers.

vii. **Impact:** The 1995 USAID evaluation report on the program noted that 'this is no longer a program, it's a movement' (quoted by Winrock International Institute for Agricultural Development, 2000).

The success of the program model resulted in being able to attract funding from a wide range of donors through unsolicited proposals, its replication in China, and proposals to take it to Nepal, Vietnam, and Latin America. Evaluation of the program revealed that 82 percent of the alumni were promoted to higher positions, while 87 percent reported increases in their salary and responsibilities. In 2006, the program transformed itself into an international organization, AWLAE-Net, with membership of professional women's associations in 10 countries – Benin, Côte d'Ivoire, Ethiopia, Ghana, Kenya, Mali, Nigeria, Senegal, Tanzania, and Uganda (Winrock International Institute for Agricultural Development, 2000).

Need for Gender Sensitisation

i. **Changing Stereotypes and Mind-sets:** Sensitization is by far the most effective and non-confrontationist approach of reforming the society. Gender sensitization in the livestock sector involves challenging traditional gender roles and stereotypes. Historically, ownership in livestock farming has often been seen with the with men as they are primarily responsible for decision-making. Women were usually relegated to tasks like milking, feeding, and cleaning.

   Example: In a rural community, a gender sensitization program can encourage men and women to work together in all aspects of livestock farming. Men can be encouraged to participate in traditionally female-dominated tasks like milking, while women can receive training and support to take on more active roles in decision-making and animal healthcare. This shift in mind set helps break down the stereotype that men and women must function in different socio-economic spaces within the livestock sector.

ii. **Increasing Sensitivity towards Women's Issues:** Gender sensitization increases the sensitivity of people at large towards women and their problems. In the process it creates a class of responsive functionaries at different level, from policy making to grass root level, who are convinced that any form of gender bias is an obstacle on the way of attaining an equitable social and economic order and therefore consider addressing gender related issues in their situation as a matter of priority. Gender sensitization aims to make people more aware of the challenges and issues that women face in the livestock sector. It highlights the importance of recognizing and addressing gender bias.

Example: Suppose there is a community where women are responsible for caring for small ruminants (such as goats) but often face challenges accessing veterinary services. Through gender sensitization, community members, including policymakers, can understand that this gender bias hinders women's participation and economic empowerment. As a result, they may prioritize providing equal access to veterinary services for women livestock keepers, acknowledging that addressing this gender-related issue is essential for achieving a more equitable social and economic order.

In summary, gender sensitization in animal husbandry refers to promoting awareness, understanding of gender issues, and ensuring gender equality in the context of animal agriculture and husbandry practices. It involves recognizing the distinct roles and contributions of men and women in the sector, addressing gender disparities, and providing equal opportunities and benefits for all genders involved in animal husbandry. It involves challenging stereotypes, promoting shared responsibilities, and increasing awareness of gender-related issues. By doing so, it can lead to a more inclusive and equitable livestock industry where both men and women have equal opportunities to participate and benefit from this economic activity.

Chapter Summary

- Gender sensitization involves raising awareness about gender equality concerns and challenging stereotypes in the context of livestock farming.
- The key components of gender sensitization for veterinarians include awareness, empathy and understanding, behaviour modification, education, and training.
- The four stages of gender sensitization are change in perception, recognition, accommodation, and action.
- Veterinary students can contribute to gender sensitization by treating male and female livestock farmers and pet owners equally, encouraging female participation, providing education and training in a gender-sensitive manner, empowering women, challenging stereotypes, offering flexible services, and collecting gender-disaggregated data.
- Gender sensitization programs can target different audiences, including men and women in livestock farming communities, elderly women, socially and economically progressive women, researchers, policy makers, and service providers.
- Content for gender sensitization programs in the livestock sector may include case studies, situation analysis, training sessions, and materials like leaflets, booklets, posters, and videos.
- Methodologies for gender sensitization in livestock farming may involve sensitization camps, mass media campaigns, theatre plays,

and participatory discussions among men and women from the same households.

- Gender sensitization is important for changing stereotypes and mind-sets in the livestock sector, as it encourages men and women to work together in all aspects of livestock farming.
- Gender sensitization increases sensitivity towards women's issues and helps address gender-related challenges and disparities in livestock farming.
- Gender sensitization in livestock production promotes awareness, understanding, and gender equality, leading to a more inclusive and equitable livestock industry.

SECTION B : SUBJECTIVE QUESTIONS

1. How can we promote gender sensitization in the veterinary services? Elaborate through an example?

Let us consider a veterinary clinic that provides medical care for animals. The staff at this clinic includes veterinarians, veterinary technicians, and administrative personnel. In this scenario, the clinic decides to implement gender sensitization initiatives to ensure equal and respectful treatment of both clients and staff, regardless of their gender.

i. **Training and Education**: The clinic organizes training sessions for all staff members on gender sensitization. These sessions focus on raising awareness about gender stereotypes, biases, and the importance of treating all clients and colleagues equally. Training can also include discussions on using inclusive language and addressing any misconceptions about specific gender roles.

ii. **Inclusive Policies**: The clinic revises its policies and procedures to ensure gender equality. This could involve implementing policies that address issues like parental leave, flexible work arrangements, and equal opportunities for career advancement for all staff members, regardless of their gender.

iii. **Client Communication**: In interactions with clients, the clinic ensures that all staff members use gender-neutral and inclusive language. Receptionists and other staff members are trained to avoid making assumptions about the roles or responsibilities of clients based on their gender.

iv. **Professional Development**: The clinic offers opportunities for professional development to all staff members equally. This could include workshops, conferences, and further education to help them excel in their roles and advance in their careers.

v. **Promoting Inclusivity**: The clinic displays posters, brochures, and information materials in the waiting area that promote gender equality and inclusivity. This helps to create a welcoming and respectful environment for all clients, regardless of their gender identity.

vi. **Feedback Mechanisms**: The clinic establishes a feedback mechanism where clients and staff members can anonymously report any instances of gender bias or insensitivity to the management. This feedback helps the clinic management continuously improve its gender sensitization efforts.

vii. **Community Engagement**: The clinic participates in community outreach programs to raise awareness about gender equality in veterinary care. This can include workshops for pet owners on responsible pet care that emphasizes equal involvement from all family members.

By implementing these gender sensitization initiatives, the veterinary clinic creates an environment where both clients and staff feel respected, valued, and equal. This, in turn, contributes to a more inclusive and just society where gender-based discrimination is actively challenged and addressed.

2. Explain how Gender Sensitisation measures can bring a change in perception in a livestock rearing community?

Imagine that in a rural community, there is a long-standing tradition that assigns specific roles within animal husbandry based on gender. Men are typically responsible for the more physically demanding tasks like feeding, cleaning, and veterinary care, while women are expected to handle tasks like milking, which are considered less strenuous.

With the introduction of gender sensitization workshops in the community, a shift in perception begins to take place. Men and women come together to learn about the importance of equality and shared responsibilities. They start to understand that assigning tasks solely based on gender perpetuates stereotypes and limits the potential for overall development.

As a result of this change in perception:

i. **Shared Responsibilities:** Men start to recognize that women are equally capable of handling tasks such as feeding and cleaning. Women, on the other hand, realize that they can contribute to physically demanding tasks like veterinary care. This leads to a more balanced distribution of responsibilities, benefiting both the livestock and the community.

ii. **Skill Development:** Men and women can learn new skills they may not have considered before. For instance, men might learn the nuances of animal nutrition from women who have been traditionally responsible for feeding. Similarly, women could benefit from the expertise of men in areas like construction of animal shelters.

iii. **Increased Efficiency:** With a broader skill set among community members, tasks are completed more efficiently. The workload is shared, reducing the burden on any one gender, and allowing everyone to focus on tasks they excel at and enjoy.

iv. **Empowerment:** As women take on roles that were previously reserved for men, they gain a sense of empowerment and independence. This empowerment extends beyond the realm of animal husbandry and influences their participation in other community decisions and activities.

v. **Changing Perceptions beyond Livestock:** The change in perception does not stop at animal husbandry. It extends to how men and women interact in various aspects of their lives, fostering a more equitable and respectful community environment.

In this hypothetical example, gender sensitization breaks down traditional gender-based barriers in animal husbandry. It encourages people to see each other as equals, capable of contributing their unique skills and strengths to the betterment of the community.

3. Explain recognition stage of gender sensitisation in a livestock rearing community?

Imagine a rural community where traditional gender roles have led to men primarily handling large animals, while women are responsible for smaller animals like poultry and goats. Gender sensitization initiatives are introduced in this community to raise awareness about the importance of recognizing women's contributions and talents in animal husbandry.

i. **Positive Qualities of Women:** As a part of the sensitization process, community members start acknowledging the positive qualities that women bring to animal husbandry. Women are known for their attention to detail, patience, and dedication in caring for smaller animals like poultry, which is crucial for maintaining the overall health and productivity of the livestock. Farm women are good at selecting fertile eggs for hatching, keeping an eye on chicks to protect them from predators etc.

ii. **Recognition of Virtues:** As the sensitization efforts continue, male members of the community begin to recognize the virtues of women's contributions to animal husbandry. They realize that the work women do with smaller animals significantly impacts the household's income and nutrition.

iii. **Appreciation for Involvement:** There is a spontaneous appreciation for the diverse role's women play in animal husbandry. Women are not only taking care of the animals' daily needs but also actively involved in activities like breeding, disease prevention, and marketing of animal products.

iv. **Increased Visibility:** With the growing recognition, women's contributions become more visible within the community. Their role in ensuring a steady supply of eggs, meat, and dairy products becomes evident, and the community starts acknowledging their efforts.

v. **Empowerment and Pride:** Women become more conscious of their capabilities and take pride in their role in animal husbandry. They gain confidence in their abilities and realize the positive impact they have on the overall well-being of the household.

vi. **Collective Efforts:** Regardless of socio-economic backgrounds, women begin to see their challenges and successes in the broader context of women's development in animal husbandry. They start acknowledging and appreciating the efforts of fellow women in different areas of livestock farming.

vii. **Recognition of Men's Role:** Through the sensitization process, women also recognize the key role that men can play in promoting gender equality in animal husbandry. Men are encouraged to share responsibilities, learn from the expertise of women, and collaborate to improve the overall productivity of the animal husbandry sector and their family welfare.

The stages of recognition in gender sensitization within the context of animal husbandry highlight the gradual shift in attitudes and perceptions towards women's contributions and the potential for gender-inclusive development.

4. Explain accommodation stage of gender sensitisation in Veterinary Practice?

In a veterinary clinic, a male veterinarian and a female veterinarian are working together on a complex case involving a sick animal. They both bring different perspectives to the table based on their experiences and knowledge. Instead of allowing any potential gender-related differences to impact their collaboration, they focus on the well-being of the animal. The male veterinarian does not dismiss the female veterinarian's suggestions but actively listens and discusses the merits of different treatment options. The female veterinarian, on her part, does not hesitate to voice her opinions and appreciates the open-mindedness of her male counterpart. Through patient and respectful dialogue, they arrive at a comprehensive treatment plan that benefits the animal's health. This example demonstrates how the principles mentioned in the text are applied in a veterinary practice setting, where both male and female professionals work together, value each other's input, and prioritize effective communication for the best outcomes.

5. Discuss three components of gender sensitization strategy in the dairy farming situation?

Imagine a rural community where dairy farming is a common livelihood. The traditional roles dictate that men are responsible for decision-making, fiscal management, and technical aspects of farming, while women primarily handle household chores and some aspects of animal care. This division of labour, although rooted in socio-cultural norms, can lead to unequal opportunities and resource allocation between male and female member within the family.

A Selecting the target audience

i. **Involving Men:** Sensitization programs would target male livestock farmers by emphasizing the benefits of gender equality in animal husbandry. For instance, workshops could highlight research showing that involving women in decision-making and management can lead to improved farm productivity and overall well-being of the animals. By involving men in these discussions, the aim is to encourage them to rethink traditional gender roles and recognize the importance of empowering women in farm operations.

ii. **Addressing Internal Barriers in Women:** The sensitization programs would also address the issue of women who inadvertently perpetuate traditional norms by conforming to them. Workshops could focus on providing information about the advantages of their active involvement in all aspects of animal care and management. For example, women try to avoid expressing their opinions or ideas in workshops/ training sessions dominated by male participants The program might also delve into ways women can overcome internalized biases and gain confidence to take on more prominent roles in the farm's decision-making processes.

iii. **Promoting Collaborative Decision-Making:** Another aspect of the program could be teaching both men and women about collaborative decision-making. By involving both genders in crucial farm decisions, it becomes possible to tap into a broader range of perspectives and experiences, leading to more informed choices and better outcomes for the farm and its animals.

B. Content of the Programme

1. Introduction

Briefly explain the importance of gender equality in all spheres of life, including animal husbandry.

Define key terms related to gender, bias, discrimination, and affirmative action.

2. Understanding Gender Biases in Animal Husbandry

Present statistics and case studies highlighting gender disparities in animal husbandry.

Discuss traditional roles assigned to men and women in the field and their impact on animal productivity and family income.

3. Exploring Women's Role in Animal Husbandry

Highlight the active participation of women in various aspects of animal husbandry, such as caregiving, milking, and management.

Share success stories of women who have excelled in animal husbandry and challenge stereotypes.

4. The Importance of Mutual Support

Emphasize the significance of cooperation between men and women in animal husbandry for better outcomes.

Discuss how shared responsibilities can lead to increased productivity and economic growth.

5. Case Studies

Present real-life examples of households or communities where gender biases have been overcome for improved animal husbandry practices. Highlight instances of men and women working together harmoniously to achieve common goals.

6. Affirmative Action and Inclusivity

Introduce the concept of affirmative action and its potential to address gender disparities.

Discuss initiatives that promote women's involvement in decision-making and leadership roles within the animal husbandry sector.

7. Debates and Discussions

Encourage participants to engage in open discussions about prevalent gender biases in animal husbandry.

Organize group debates on topics like "The Role of Women in Livestock Management" or "Creating Gender-Responsive Animal Husbandry Practices."

8. **Addressing Concerns and Myths**

   Address common concerns and myths that perpetuate gender biases in animal husbandry.

   Provide evidence-based information to counter the prevailing misconceptions and stereotypes.

9. **Interactive Activities**

   Conduct interactive activities that challenge participants' perceptions of gender roles and responsibilities.

   Role-playing scenarios can help participants empathize with challenges faced by individuals in distinct roles.

10. **Creating an Action Plan**

    Facilitate a session where participants collectively brainstorm strategies to promote gender equality in their own animal husbandry practices.

    Encourage participants to commit to implementing minor changes that can contribute to a more inclusive environment.

11. **Conclusion**

    Summarize key takeaways from the program.

    Reinforce the message that gender equality is essential for sustainable development in animal husbandry and beyond.

    Remember that the content should be adapted to the specific cultural context and audience preferences. The goal is to spark thoughtful discussions, challenge preconceptions, and inspire positive changes in attitudes and behaviours towards gender roles and biases in animal husbandry.

C. Methodology

Gender-Sensitive Modules

Develop modules that highlight gender dynamics in animal husbandry. These modules could include case studies highlighting successful experiences of women in livestock management, as well as situation analyses that highlight the challenges they face.

Gender-Sensitive Materials

Create gender-sensitive educational materials such as leaflets, booklets, posters, and videos. For instance, a video could feature women from diverse backgrounds sharing their innovative practices in animal care, demonstrating that women can excel in traditionally male-dominated roles.

Sensitization Camps

Organize sensitization camps in rural areas where women are actively involved in animal husbandry. These camps would include hands-on training, discussions on gender roles, and practical demonstrations of improved livestock management techniques. The camps would encourage women to take a more active role in decision-making and leadership within their communities.

Mass Media Campaign

Launch a sustained campaign through mass media channels such as radio, TV, and newspapers, social media etc. Develop stories that challenge gender stereotypes in animal husbandry. For instance, highlight male farmers who actively involve their wives in decision-making and emphasize the positive impact on their farms' productivity.

Participatory Discussions

Facilitate participatory discussions involving men and women of different age groups from the same households. Create an enabling environment for open conversations about gender biases and their real-life implications. Sharing firsthand experiences can help participants recognize the adverse effects of gender bias and promote change.

Expected Outcomes

Increased Women's Participation: By promoting gender-sensitive practices and providing opportunities for women's involvement, there should be an increase in the active participation of women in livestock management and decision-making.

Improved Livestock Practices

The dissemination of gender-sensitive materials and training modules would contribute to the adoption of improved livestock management practices, contributing to both productivity and animal welfare.

Enhanced Community Awareness

The mass media campaign and sensitization camps would raise awareness about the importance of gender equity in animal husbandry, encouraging a more inclusive and supportive community.

Empowered Women

Through participatory discussions, women would gain a stronger voice in their households and communities, leading to increased confidence and empowerment.

Sustainable Change

By addressing gender biases at various levels and involving both men and women, the changes are more likely to be sustained eventually.

6. Why Agricultural Extension and advisory services must be Gender Sensitive?

Agricultural extension and advisory services must be gender-sensitive for several compelling reasons. Firstly, recognizing the pivotal role of agricultural growth as a means of poverty alleviation underscores the need to enhance the productivity of women farmers. Given that women constitute most farmers and laborers in numerous countries (GDPRD, 2010), fostering their active participation in agricultural activities is crucial for overall sector development.

Addressing gender inequality is not just a moral imperative; it is an economic necessity. Investing in efforts to dismantle barriers faced by women in agriculture has been identified as a key strategy to boost efficiency and productivity in the agricultural sector. A report by FAO argues that if women were to have access to the same level of resources as men, agricultural productivity would increase by up to 30%, agricultural output by up to 4%, and the number of poor people would reduce by up to 17%. The potential for such initiatives to contribute to overall economic growth and poverty reduction has been acknowledged by prominent organizations like the World Bank (2007), Christoplos (2010), and World Bank/FAO/IFAD (2009).

The awareness of the need for gender sensitivity in agricultural extension and advisory services emerged in the early 1980s, coinciding with the realization of women's significant contributions to agriculture. As women and men often assume distinct roles and responsibilities in agricultural activities, it became evident that advisory services must adapt to the specific needs and demands of each gender. However, a lack of awareness among advisors and staff regarding these gender dynamics necessitated capacity-building efforts, including training on "gender analysis" and "gender-sensitive agricultural planning."

The gender gap in access to crucial resources, such as assets, information, markets, and credit, became apparent, emphasizing the urgency of addressing these disparities. Women's exclusion from advisory services due to their often-overlooked status as "legitimate clients" further underscored the need for a gender-sensitive approach (www.reachingruralwomen.org).

Agricultural extension and advisory services play a pivotal role in facilitating interaction, knowledge sharing, and the adoption of modern technologies among farmers. However, despite their importance, access to extension services remains low for both women and men in developing economies, with women underutilizing these services compared to men (Meinzen-Dick *et al.*,

2010). Closing this gender gap in access to extension services is essential for realizing the full potential of agricultural development and ensuring that both women and men benefit equitably from these crucial resources.

7. How can learning within groups empower women producers, and what proactive measures are suggested for enhancing women's participation in producer organizations?

Gender-sensitive agricultural extension learning within groups empowers women producers by providing opportunities to strengthen their articulation and bargaining power with authorities, service providers, and policymakers. Proactive measures include setting up quotas for women or involving members' spouses in producer organizations. Organizing seminars exclusively for women, facilitated by female extension advisers, is a valuable strategy to address gender-specific challenges. The diverse needs of women, spanning agricultural production, food processing, nutrition, household tasks, child education, and family health, highlight the importance of making multiple services available, affordable, and accessible. In project design, adopting a gender-sensitive approach serves as an entry point for promoting equitable participation. During implementation, substantiating and adapting to the diverse needs and demands of both men and women is crucial. Collecting feedback and evaluating project activities from both genders contribute to improved project performance and impact, ensuring responsiveness to evolving community needs for more sustainable outcomes.

Summary

- The chapter focuses on the concept of gender sensitization in the context of livestock farming, which aims to promote awareness of gender equality concerns and challenge traditional gender roles and stereotypes.
- Gender sensitization involves modifying behaviour by raising awareness of gender equality issues and encouraging behaviour change.
- It is essential to target different audiences in livestock farming communities, including men, women, and elderly women, to address and eliminate gender bias.
- The content of gender sensitization programs should be easily understandable and may include case studies, situation analysis, educational materials, and community discussions.
- Methodologies for gender sensitization can include gender-sensitive modules, materials, workshops, sensitization camps, mass media campaigns, plays, and participatory discussions.

- Gender sensitization is necessary for changing stereotypes and mindsets, encouraging men and women to work together in all aspects of livestock farming.
- It increases sensitivity towards women's issues in the livestock sector and highlights the importance of addressing gender bias.
- Gender sensitization aims to achieve gender equality and promote shared responsibilities in animal husbandry, fostering inclusivity and cooperation in the industry.

SECTION C: SHORT ANALYTICAL QUESTIONS

1. **What are the key components of a gender sensitization strategy in the context of livestock farming, and how can these components be applied to promote gender equality in the sector?**

   The key components of a gender sensitization strategy in livestock farming include selecting the target audience, deciding the content, and determining the methodology. Target audiences can range from men and women in livestock farming communities to elderly women and socially progressive women. The content should focus on educating about gender biases and promoting women's participation. The methodology can involve gender-sensitive modules, materials, sensitization camps, media campaigns, and participatory discussions. By implementing these components, gender sensitization can promote gender equality in the sector.

2. **How does gender sensitization help change stereotypes and mindsets in the livestock sector, and what are the benefits of challenging traditional gender roles in this context?**

   Gender sensitization challenges traditional gender roles and stereotypes in the livestock sector by encouraging men and women to work together in all aspects of livestock farming. It helps break down the stereotype that men and women have distinct roles within the sector. By doing so, it increases cooperation and inclusivity, allowing individuals to choose tasks based on their skills rather than their gender. This shift in mindset leads to a more equitable and productive livestock industry, where all stakeholders can benefit from shared responsibilities and decision-making.

3. **How does gender sensitization increase sensitivity towards women's issues in livestock farming, and why is addressing gender-related problems essential for achieving equitable social and economic orders?**

   Gender sensitization increases sensitivity towards women's issues in livestock farming by making people more aware of the challenges and problems that women face in the sector. It highlights the importance of recognizing and addressing gender bias and disparities. Addressing gender-related issues is essential for achieving equitable social and economic orders because it ensures that all genders have equal opportunities and benefits. When gender bias hinders women's participation and economic empowerment, addressing these issues becomes a priority to achieve a more equitable and inclusive livestock sector and a more equitable society overall.

4. **What is the definition of gender sensitization in the context of livestock farming?**

   Gender sensitization in livestock farming involves raising awareness about gender equality concerns, challenging stereotypes, and promoting equitable practices within the sector.

5. **What are the stages of gender sensitization, and how do they apply to the livestock industry?**

   Gender sensitization generally involves four stages: change in perception, recognition, accommodation, and action. In livestock farming, these stages can be seen in changing traditional gender roles, recognizing the contributions of both men and women, breaking down gender barriers, and taking affirmative actions to promote women's participation.

6. **How can veterinary students contribute to gender sensitization in the livestock sector?**

   Veterinary students can contribute by treating male and female livestock farmers and pet owners equally, encouraging female participation, providing education and training in a gender-sensitive manner, empowering women with essential knowledge and skills, challenging gender stereotypes, offering flexible services, and collecting gender-disaggregated data related to animal health and management.

7. **What is the target audience for gender sensitization in the livestock sector, and how can they be sensitized?**

   The target audience includes men and women in livestock farming communities, elderly women who uphold traditional gender roles, socially and economically progressive women, researchers, policy makers, and service providers. Sensitization can be achieved through gender-sensitive modules, educational materials, sensitization camps, mass media campaigns, theater plays, and participatory discussions.

8. **Why is gender sensitization important in the livestock sector, and what are some examples of its benefits?**

   Gender sensitization is important as it challenges stereotypes, increases sensitivity towards women's issues, and fosters a more inclusive and equitable livestock industry. For example, it can lead to men and women working together in all aspects of livestock farming, increase sensitivity towards women's issues such as accessing veterinary services, and ultimately promote a more equitable social and economic order in the sector.

9. **How does gender sensitization contribute to breaking stereotypes and mindsets in the livestock sector?**

Gender sensitization challenges traditional gender roles and stereotypes by encouraging men and women to work together in all aspects of livestock farming. It involves breaking down the stereotype that men and women should function in different socio-economic spaces within the livestock sector

10. **Why is gender sensitization essential in the livestock industry?**

Gender sensitization is essential in the livestock industry to challenge traditional gender roles, break down stereotypes, and foster inclusivity and cooperation among all stakeholders, ultimately leading to increased productivity, sustainability, and well-being.

11. **What are the stages of gender sensitization, and how do they lead to behaviour modification?**

The stages of gender sensitization include change in perception, recognition, accommodation, and action. These stages lead to behaviour modification by changing individuals' views, recognizing the value of each gender's contributions, accommodating each other's roles, and taking action to promote gender equality.

12. **How can gender sensitization benefit the livestock industry?**

Gender sensitization can benefit the livestock industry by enhancing productivity, sustainability, and cooperation. It helps challenge traditional gender roles, leading to more inclusive practices and improved animal welfare.

13. **What are the key steps in gender sensitization for veterinarians?**

The key steps in gender sensitization for veterinarians involve creating awareness about existing gender biases, fostering empathy and understanding, actively modifying behavior to promote gender equality, and providing education and training on gender sensitization.

14. **How can veterinarians promote gender equality and fairness in their profession?**

Veterinarians can promote gender equality by actively working to modify behavior, encouraging both male and female colleagues to participate in tasks traditionally associated with the opposite gender, and providing training and workshops on gender sensitization.

15. **What are some key considerations for veterinary students regarding gender sensitization?**

    Veterinary students should consider equal treatment of male and female livestock farmers and pet owners, encouraging female participation, providing gender-sensitive education and training, empowering women, challenging stereotypes, offering flexible services, and collecting gender-disaggregated data.

16. **How can veterinary students encourage equal treatment of male and female livestock farmers and pet owners?**

    Veterinary students can treat both genders equally by giving equal weight to their concerns, opinions, and knowledge about their animals and ensuring that everyone's voice is heard in discussions and decision-making.

17. **Why is it essential for veterinary students to empower women in the livestock industry?**

    Empowering women in the livestock industry can lead to improved animal care and overall community development, benefiting both male and female livestock farmers and pet owners.

18. **How can gender sensitization benefit the livestock industry's approach to animal care and agriculture?**

    Gender sensitization can contribute to a more gender-sensitive and equitable approach to animal care, fostering positive relationships within the community, and improving the well-being of animals in their care.

19. **What is the significance of recognizing and appreciating the roles of women in livestock farming and pet ownership?**

    Recognizing and appreciating the roles of women in livestock farming and pet ownership helps promote gender sensitivity and equality in the industry, leading to better animal care and community development.

20. **How can veterinary students foster collaboration with women's groups and organizations in the livestock sector?**

    Veterinary students can collaborate with women's groups and organizations to enhance knowledge exchange and implement joint initiatives aimed at improving animal health and welfare.

21. **What are the key components of a gender sensitization strategy?**

    A gender sensitization strategy includes selecting the target audience, deciding the content, and determining the methodology for raising awareness about gender equality.

22. **Why is it important to target both men and women in gender sensitization programs in livestock farming communities?**

    Targeting both men and women ensures that everyone is aware of the benefits of involving women in livestock management, promotes gender equality, and increases overall productivity and sustainability in the sector.

23. **How can gender sensitization programs educate elderly women about gender bias in livestock farming?**

    Gender sensitization programs can educate elderly women about the negative consequences of gender bias by organizing community discussions and storytelling sessions where they can learn from successful cases of gender-inclusive livestock management.

24. **What role can socially and economically progressive women play in gender sensitization programs in the livestock sector?**

    Socially and economically progressive women can serve as role models and mentors, supporting and empowering underprivileged women in the livestock sector and helping them access training and resources.

25. **How can gender sensitivity be promoted through materials like leaflets, booklets, posters, and videos in the livestock sector?**

    These materials can be designed to educate people about gender equity by challenging stereotypes and showcasing successful stories of gender equality and shared responsibilities in animal care.

26. **Why is it essential to use participatory discussions among men and women of different age groups in gender sensitization programs?**

    Participatory discussions create a platform for sharing practical experiences and understanding the challenges women face in livestock farming. They foster empathy and cooperation among different generations in the community.

27. **How does gender sensitization help in changing stereotypes and mind-sets in the livestock industry?**

    Gender sensitization challenges traditional gender roles and stereotypes in the livestock industry, encouraging both men and women to work together and participate in various aspects of livestock farming.

28. **What are the benefits of increasing sensitivity toward women's issues through gender sensitization?**

    Increasing sensitivity toward women's issues through gender sensitization leads to a better understanding of the challenges and issues women face in the livestock sector, ultimately promoting gender equity.

29. Why is it crucial to change mind-sets in the livestock sector regarding traditional gender roles?

Changing mindsets regarding traditional gender roles is essential to promote inclusivity, cooperation, and equal opportunities in the livestock industry, leading to increased productivity and sustainability.

30. How can gender sensitization programs create awareness about the ill effects of gender bias and discriminatory practices in the livestock sector?

Gender sensitization programs can create awareness by initiating friendly debates, conducting case studies, and discussing the conduct of men and women in households based on real-life scenarios. This encourages people to address gender bias.

31. What are the implications of promoting gender sensitivity on individual and societal levels?

Promoting gender sensitivity at both individual and societal levels helps create a responsive environment for addressing gender-related issues and achieving an equitable social and economic order.

32. How can gender sensitization impact the roles and responsibilities of men and women in the livestock sector?

Gender sensitization can impact the roles and responsibilities of men and women by encouraging a shared approach to managing livestock and challenging stereotypes and biases.

33. What are some examples of actions that can improve women's conditions in the livestock sector?

Actions to improve women's conditions might include creating financial assistance programs for women starting veterinary practices, establishing community veterinary clinics accessible to women, and conducting research to enhance livestock breeding techniques suitable for women farmers.

34. How does gender sensitization influence the conduct of men and women within a household in the livestock sector?

Gender sensitization encourages both men and women to work together in the livestock sector, share responsibilities, and understand the benefits of cooperative participation for the well-being of the family and community.

35 Why is mass media an effective tool for spreading awareness about gender sensitivity in the livestock sector?

Mass media can reach a wide audience and challenge stereotypes by featuring advertisements, programs, and plays that promote gender equality and share stories of successful gender-inclusive practices.

36. How can gender sensitization lead to an equitable and productive livestock industry?

Gender sensitization can lead to an equitable and productive livestock industry by fostering positive relationships, promoting shared responsibilities, and increasing awareness of gender-related issues, ultimately benefiting the entire sector.

37. What are the three key dimensions to consider when applying gender mainstreaming, and why are they important?

The three key dimensions are the technical dimension, political dimension, and cultural dimension. They are important because they provide a holistic approach to implementing gender mainstreaming, taking into account data, political advocacy, and cultural context.

38. How does gender analysis contribute to the effective implementation of gender mainstreaming at the programmatic level?

Gender analysis helps identify how and why certain issues affect women and men differently, allowing for the development of targeted solutions and program design that addresses gender disparities.

39. Why is it essential to allocate adequate resources for gender mainstreaming initiatives within institutions?

Adequate resources are necessary to support gender mainstreaming, including conducting gender analysis, developing gender-responsive programs, and providing gender-focused training and capacity-building activities.

40. What are the potential benefits of gender mainstreaming in terms of social and economic development?

Gender mainstreaming can contribute to increased economic growth and improved social indicators by utilizing the skills and talents of all individuals, regardless of their gender.

41. How can gender mainstreaming help reduce gender-based violence?

Gender mainstreaming challenges unequal power dynamics and traditional gender norms, which can contribute to the reduction of gender-based violence.

42. How does the availability of reliable and sex-disaggregated data contribute to the effectiveness of gender mainstreaming in practice?

Reliable and sex-disaggregated data are essential for gender mainstreaming as they provide a factual basis for understanding gender disparities, formulating targeted interventions, and measuring progress. When such data is available, it becomes easier to identify areas where men and women are impacted differently, design programs to address these disparities, and track changes over time to ensure gender equality.

43. What is the significance of establishing an empowering environment for women to assess their circumstances and voice their preferences in the context of gender mainstreaming?

Establishing an empowering environment for women to assess their circumstances and voice their preferences is critical for gender mainstreaming. It ensures that women are not just passive recipients of development initiatives but active participants in defining their needs and priorities. This approach recognizes that women are best positioned to identify the challenges they face and the solutions that will work for them, leading to more effective and sustainable development outcomes.

44. How do socio-cultural norms and local ideologies influence the execution and interpretation of gender mainstreaming initiatives, and why is it important to consider these factors?

Socio-cultural norms and local ideologies play a significant role in shaping the success of gender mainstreaming initiatives. It's important to consider these factors because they can either support or hinder the implementation of gender equality strategies. Understanding local norms and ideologies helps in designing interventions that are culturally sensitive and contextually relevant. Failure to do so can lead to resistance and the ineffectiveness of gender mainstreaming efforts.

45. What are the key steps involved in gender mainstreaming at the programmatic level, and how do these steps contribute to promoting gender equality?

The key steps in gender mainstreaming at the programmatic level include gender analysis, program design, resource allocation, implementation, and monitoring and evaluation. These steps contribute to promoting gender equality by ensuring that programs are tailored to address the specific needs and challenges faced by different genders, that resources are allocated to support these goals, and that progress is measured and adjustments are made as necessary to achieve gender equality outcomes.

46. How can the establishment of institutional structures, resource allocation, and accountability mechanisms contribute to the successful incorporation of gender considerations into institutions?

The establishment of institutional structures, resource allocation, and accountability mechanisms are essential for the successful incorporation of gender considerations into institutions. Institutional structures like gender units provide the necessary framework for gender mainstreaming efforts, while resource allocation ensures that there are adequate funds and personnel dedicated to gender-related initiatives. Accountability mechanisms, such as gender-sensitive indicators, help institutions track and measure progress in achieving gender equality goals, ensuring that these considerations are integrated into the institution's core functions.

47. What are the key drivers for the need for gender mainstreaming in society, and how do they impact various aspects of development and social change?

The key drivers for the need for gender mainstreaming in society include the promotion of gender equality, addressing gender disparities, enhancing policy and program effectiveness, fostering social and economic development, reducing gender-based violence, ensuring human rights, fulfilling international commitments, improving health outcomes, empowering marginalized groups, breaking stereotypes, and achieving long-term social change. These drivers impact various aspects of development and social change by promoting inclusivity, equity, and justice, and by challenging historical and structural inequalities that have affected different genders.

48. What is the fundamental shift in objectives in gender-sensitive livestock extension advisory services, and how does it impact socio-economic well-being?

The shift is from merely enhancing production to facilitating improved income and productive employment opportunities for both genders. This change acknowledges the broader impact of veterinary and animal husbandry practices on the socio-economic landscape. It ensures sustainable and inclusive growth, emphasizing equitable economic opportunities and improved livelihoods for both men and women. This evolution makes livestock extension more holistic and gender-sensitive.

49. How does the transition from self-help groups to common-activity groups contribute to inclusivity in veterinary and animal husbandry activities?

The shift encourages collaborative engagement by establishing common-activity groups open to individuals of all genders. This ensures shared responsibilities, promoting inclusivity and equitable participation for both men and women in veterinary and animal husbandry activities.

50. What is the significance of the change from input distribution to developing local capacity in gender-sensitive livestock extension services?

The emphasis on developing local capacity ensures sustainable availability and management of inputs and services. This shift promotes skills and knowledge within the community, reducing dependency on external input distribution and contributing to the long-term success of veterinary and animal husbandry extension.

51. How does the shift to a demand-led approach in intervention selection contribute to a more gender-sensitive extension service?

The shift involves analysing client data disaggregated by gender, aligning interventions with identified opportunities, and ensuring the availability of complementary support and services. This approach makes interventions more responsive to the specific needs of men and women, contributing to a more gender-sensitive agricultural and animal husbandry extension.

52. In what ways does the evolving, diverse, and gender-differentiated approach in working with women enhance the effectiveness of animal husbandry extension services?

Answer: The approach involves customizing extension services based on the unique requirements and roles of men and women in animal husbandry. This ensures that services are tailored to address specific challenges and opportunities faced by men and women, making the extension more relevant, effective, and responsive to the diverse needs of the community.

53. How does the shift from fixed approaches to working with women to a partnership-based approach contribute to more inclusive decision-making in animal husbandry practices?

The shift involves recognizing women as active contributors to decision-making processes related to animal husbandry. This can be achieved by establishing women's cooperatives and involving women in community-level committees. This ensures that women's perspectives are considered in decisions related to disease control programs, vaccination schedules, and overall animal husbandry practices.

54. What is the significance of shifting from input and output targets to behavioural and livelihood changes in gender-sensitive monitoring and evaluation?

Answer: This shift assesses not only quantitative outputs but also the behavioral and livelihood changes resulting from extension services. It evaluates changes in livestock health, milk production, income levels, and involvement of women and youth in decision-making. This broader evaluation provides a more comprehensive understanding of the impact of veterinary extension programs.

55. How does gender sensitization contribute to breaking stereotypes and mindsets in the livestock sector?

Answer: Gender sensitization challenges traditional gender roles and stereotypes by encouraging men and women to work together in all aspects of livestock farming. It involves breaking down the stereotype that men and women should function in different socio-economic spaces within the livestock sector.

56. **What is the role of gender sensitization in increasing awareness of women's issues in the livestock sector?**

    Answer: Gender sensitization increases the sensitivity of people towards the challenges and issues that women face in the livestock sector. It aims to make people more aware of gender bias and emphasizes the importance of recognizing and addressing gender-related issues for achieving a more equitable social and economic order.

57. **How can gender sensitization lead to more inclusive and equitable livestock industry practices?**

    Gender sensitization involves promoting awareness and understanding of gender issues, recognizing different roles and contributions, and addressing gender disparities in animal agriculture. By challenging stereotypes, promoting shared responsibilities, and increasing awareness, gender sensitization contributes to a more inclusive and equitable livestock industry where both men and women have equal opportunities to participate and benefit.

SECTION D: SUBJECTIVE QUESTIONS

Question 1. Multiple Choice Questions

1. **What is the primary goal of gender sensitization?**

    a. Encourage stereotypes b. Challenge traditional roles

    c. Maintain gender bias d. Promote gender inequality

2. **According to the information, what is one aspect of gender sensitization for veterinarians?**

    a. Stereotyping b. Empathy and understanding

    c. Gender bias reinforcement d. Gender segregation

3. **Gender sensitization aims to:**

    a. Reinforce traditional gender roles

    b. Create inequality between genders

    c. Address gender disparities and promote equality

    d. Promote gender bias

4. **What is the primary goal of gender sensitization?**

    a. Promoting gender bias

    b. Encouraging stereotypes

c. Modifyingbehaviour through gender equality awareness

d. Suppressing awareness of gender concerns

5. Which stage of gender sensitization involves breaking down stereotypes and recognizing the virtues of women?

a. Change in perception b. Accommodation

c. Empathy and Understanding d. Recognition

6. What is one strategy to promote gender sensitivity among veterinary students?

a. Treating male and female livestock farmers differently

b. Challenging stereotypes

c. Ignoring the concerns of female pet owners

d. Avoiding dialogue between male and female farmers

7. What is gender sensitization primarily concerned with?

a. Modifying animal behaviour

b. Modifying human behaviour through awareness of gender equality concerns

c. Promoting gender bias

d. Encouraging stereotypes

8. According to the text, gender sensitization encourages individuals to do what?

a. Ignore personal attitudes and beliefs

b. Embrace stereotypes about gender

c. Challenge and change their behaviour

d. Avoid questioning their beliefs

9. What is the first stage of the gender sensitization process?

a. Action b. Accommodation

c. Recognition d. Change in perception

10. In the livestock sector, gender sensitization might involve promoting the understanding of:

a. Traditional gender roles

b. Increased gender bias

c. Roles and contributions of both men and women

d. Stereotypes about men's and women's capabilities

11. Gender sensitization aims to:

a. Promote gender bias

b. Challenge traditional gender roles and stereotypes

c. Maintain the status quo

d. Avoid discussing gender issues

12. What is the primary purpose of gender sensitization in the livestock sector?

a. To reinforce traditional gender roles

b. To discourage women from participating in animal care

c. To promote gender equality and inclusivity

d. To create more gender bias

13. Change in perception, recognition, accommodation, and action are the stages in:

a. Animal healthcare

b. Gender sensitization

c. Livestock management

d. Veterinary education

14. Sensitization programs in the livestock sector should target:

a. Only women in rural communities

b. Both men and women involved in livestock farming

c. Elderly women in rural areas

d. Young women interested in animal husbandry

15. Which of the following is an example of the "Action" stage in gender sensitization?

a. Promoting traditional gender roles

b. Organizing workshops for women to discuss their concerns

c. Encouraging men and women to maintain gender bias

d. Ignoring gender-related issues

16. Gender sensitization modules may contain:

a. Recipes for livestock feed

b. Case studies and situation analyses

c. Folk songs about gender roles

d. Jokes about traditional gender stereotypes

17. Gender sensitization materials in the livestock sector might include:

a. Candy bars
b. Leaflets, booklets, posters, and videos
c. Toys for farm animals
d. Educational board games

18. Gender Sensitization camps could help participants:

a. Reinforce traditional gender roles
b. Engage in heated debates
c. Understand and discuss gender issues
d. Promote gender bias

19. The need for gender sensitization in the livestock sector arises from:

a. A desire to maintain gender bias
b. Changing stereotypes and Mindsets
c. The goal of reinforcing traditional gender roles
d. Avoiding discussions about gender issues

20. Gender sensitization encourages the recognition of the importance of:

a. Gender bias
b. Shared responsibilities between men and women
c. Maintaining traditional gender roles
d. Gender stereotypes

21. What is the primary goal of gender sensitization?

a. To promote gender bias
b. To maintain traditional gender roles
c. To raise awareness of gender equality concerns
d. To reinforce stereotypes

22. In the context of gender sensitization for veterinarians, what is the role of empathy and understanding?

a. To reinforce gender stereotypes
b. To encourage behaviour modification
c. To maintain gender bias
d. To reinforce gender roles

23. **Which stage of gender sensitization involves changing one's perception and seeing women asequal partners in socio-economic development?**

a. Change in perception b. Recognition
c. Accommodation d. Action

24. **What is one of the key points to ponder for veterinary students in promoting gendersensitization?**

a. Treating male and female livestock farmers differently
b. Encouraging gender bias
c. Challenging stereotypes
d. Reinforcing traditional gender roles

25. **Which component of gender sensitization strategy focuses on the selection of the targetaudience?**

a. Content b. Methodology
c. Target audience d. None of the above

26. **What is the purpose of gender-sensitive materials in gender sensitization programs?**

a. To perpetuate gender bias
b. To reinforce stereotypes
c. To educate people about gender equity
d. To maintain traditional gender roles

27. **Which stage of gender sensitization involves recognizing the virtues and importance of women in society?**

a. Change in perception b. Recognition
c. Accommodation d. Action

28. **Why is gender sensitization important in livestock farming?**

a. To reinforce traditional gender roles
b. To challenge stereotypes and improve the sector's productivity
c. To maintain gender bias
d. To promote gender inequality

29. **What can gender sensitization help individuals achieve?**

a. Maintain traditional gender roles
b. Challenge stereotypes and question gender biases

c. Promote gender inequality
d. Perpetuate gender bias

30. What is gender sensitization in animal husbandry?

a. Breeding practices
b. Promoting gender equality and awareness
c. Feeding livestock
d. Managing farm finances

31. Which of the following is not a stage in gender sensitization?

a. Change in perception
b. Empathy and understanding
c. Recognition
d. Isolation

32. In the livestock sector, what does gender sensitization primarily aim to challenge?

a. Traditional gender roles and stereotypes
b. Animal health issues
c. Government policies
d. Market prices

33. What is the first step in the gender sensitization process in livestock farming communities?

a. Promoting women's involvement
b. Recognizing virtues of women
c. Changing the behaviour of livestock
d. Enhancing men's participation

34. What are the primary targets of gender sensitization programs in livestock farming communities?

a. Only women
b. Both men and women
c. Only the elderly
d. Government officials

35. What is one benefit of increasing gender sensitivity in livestock farming communities?

a. Decreased animal productivity
b. Lower household income
c. Improved livestock management
d. Gender-based disputes

36. Which methodology could be used in gender sensitization programs for livestock communities?

a. Mass media campaigns
b. Individualcounselling only
c. Exclusive focus on men
d. Promoting stereotypes

37. In the context of livestock farming, what does "affirmative action" primarily refer to?

a. Avoiding any action related to gender sensitization
b. Taking steps to address gender bias and promote equality
c. Promoting gender stereotypes
d. Separating men and women in the sector

38. What is the ultimate goal of gender sensitization in animal husbandry?

a. To maintain traditional gender roles
b. To achieve gender equality and improve livestock practices
c. To isolate men and women in the industry
d. To promote gender bias

39. In animal husbandry, what is a key aspect of behaviour modification for gender sensitization?

a. Reinforcing traditional gender roles
b. Encouraging men to focus on large animals and women on small animals
c. Promoting shared responsibilities and challenging stereotypes
d. None of the above

40. How can veterinary students promote gender sensitization in their work?

a. Treating male and female clients differently
b. Exclusively working with one gender for client convenience
c. Tailoring advice to the specific needs of clients regardless of gender
d. None of the above

41. What is the role of the "Action" stage in gender sensitization?

a. It focuses on reinforcing gender stereotypes.
b. It encourages individuals to work toward gender equality and create a favourable environment for women.
c. It maintains the status quo of unequal gender relations.
d. None of the above

42. How can gender sensitization be achieved through mass media campaigns in the livestocksector?

a. By promoting traditional gender roles
b. By showcasing successful stories of gender equality

c. By reinforcing stereotypes

d. None of the above

43. What is the purpose of sensitization camps in gender sensitization programs?

a. To promote traditional gender norms

b. To segregate men and women based on their livestock responsibilities

c. To provide a platform for open dialogue and understanding of gender issues

d. None of the above

44. What is one way to encourage women's participation in livestock management in a sensitizationprogram?

a. Discourage women from participating in decision-making.

b. Provide training and resources tailored to women's needs.

c. Promote gender bias within the community.

d. None of the above

45. What is the first stage in the process of gender sensitization?

a. Recognition

b. Accommodation

c. Awareness

d. Action

46. In the context of veterinary care, how can gender sensitization contribute to behaviour modification?

a. By reinforcing stereotypes related to male and female veterinarians

b. By encouraging male veterinarians to handle only large animals

c. By actively working towards behaviour modification to promote gender equality

d. By excluding female veterinarians from professional opportunities

47. Which group should gender sensitization programs target in livestock farming communities topromote gender equality?

a. Only elderly women

b. Only young women

c. Both men and women

d. Only male livestock farmers

48. What does "Accommodation" represent in the stages of gender sensitization?

a. Recognizing the importance of women's contributions

b. Exercising patience and restraint

c. Overcoming traditional barriers

d. Changing perceptions about gender

49. How can men and women work together to promote gender equality in livestock farming?

a. By reinforcing traditional gender roles

b. By challenging stereotypes and sharing responsibilities

c. By excluding the opposite gender from decision-making processes

d. By avoiding cooperation in the livestock sector

50. What role can progressive women in a community play in promoting gender sensitivity inlivestock farming?

a. They should ignore the issues faced by underprivileged women.

b. They can serve as role models and mentors to empower underprivileged women.

c. They should not get involved in gender-related discussions.

d. They should isolate themselves from women's cooperatives.

51. What does "Change in Perception" represent in the stages of gender sensitization?

a. Recognizing the importance of women's contributions

b. Overcoming traditional barriers

c. Exercising patience and restraint

d. Changing perceptions about gender

52. Why is gender sensitization important in the livestock sector?

a. To reinforce traditional gender roles

b. To exclude women from the livestock industry

c. To promote gender equality and enhance productivity

d. To maintain gender bias in animal husbandry

Question 2. True or False

1. Gender sensitization involves changing behaviour through awareness of gender equality concerns.
2. Gender sensitization focuses on instilling empathy for one's own and the other sex.
3. Gender sensitization primarily targets only one gender for behaviour modification.

4. Gender sensitization can help individuals examine their personal attitudes and beliefs about gender.
5. One stage of gender sensitization involves recognizing women's virtues and contributions.
6. Gender sensitization can lead to an increased awareness of the role of women in the family and society.
7. The behaviour modification stage of gender sensitization involves accommodating gender biases.
8. Gender sensitization encourages the removal of gender barriers in professional fields.
9. Gender sensitization in veterinary medicine mainly focuses on the needs of male veterinarians.
10. Veterinarians can help break down gender stereotypes by actively communicating with male & female pet owners.
11. Gender sensitization only targets livestock farmers and excludes pet owners.
12. Training sessions on gender sensitization arc primarily for male veterinary students.
13. Gender sensitization promotes flexible veterinary services that accommodate various schedules.
14. Gender disaggregated data can help identify gender-specific challenges in animal care.
15. Gender sensitization aims to segregate men and women in the workforce.
16. Gender sensitization encourages men to resist change and be less flexible.
17. Empowering women in veterinary sciences is a key element of gender sensitization.
18. Gender sensitization involves promoting dialogue to resolve disputes.
19. Gender sensitization programs should not target elderly women.
20. Providing financial assistance to aspiring female veterinarians is an affirmative action.
21. Gender-sensitive materials include pamphlets and brochures.
22. Sensitization camps can help engage the community in gender equality discussions.
23. Sensitization programs focus solely on underprivileged women.

24. Gender sensitivity promotes the concept of shared responsibility.
25. Gender sensitization encourages elderly women to uphold traditional gender roles.
26. Participatory discussions should involve only members of the same gender within a household.
27. Affirmative action's aim to restrict women's access to resources.
28. Creating women-centric cooperatives may provide resources and financial services to women.
29. Sensitization programs only target economically progressive women.
30. Gender sensitization in livestock farming may focus on improving milk production.
31. Theatre plays have no role in gender sensitization.
32. Promoting gender sensitivity can enhance the productivity of livestock farming.
33. Sensitization can break down traditional gender roles in livestock farming.
34. Gender sensitivity aims to promote unequal distribution of livestock-related tasks.
35. Gender sensitization primarily focuses on encouraging men to be more involved in traditional female tasks.
36. Gender sensitization does not challenge traditional gender roles in rural communities.
37. Gender sensitization increases awareness of women's issues.
38. Gender sensitization has no impact on policymaking in the livestock sector.
39. Sensitization programs aim to create a class of non-responsive functionaries.
40. Creating awareness about the benefits of involving women in livestock management is a goal of gender sensitization.
41. Gender sensitivity encourages the recognition of the negative consequences of gender bias.
42. Gender sensitization primarily targets the livestock community and excludes policymakers.
43. Sensitization modules do not include case studies.

44. Gender sensitization materials should be complex and difficult to understand.
45. Sensitization programs do not focus on content designed to address the specific needs of the community.
46. Sensitization content should initiate friendly debate among the audience.
47. The methodology of gender sensitization does not involve participatory discussions.
48. Promoting gender sensitivity in animal husbandry mainly targets female roles in livestock farming.
49. Gender sensitization primarily benefits one gender at the expense of the other.
50. Gender sensitization primarily reinforces traditional gender roles.
51. In the "Action" stage of gender sensitization, individuals who have been sensitized can play a pivotal role in improving the status of women in society.
52. Sensitization programs in the livestock sector should only target women.
53. Gender sensitization aims to increase sensitivity to women's issues.
54. Gender-sensitive modules often contain complex scientific data.
55. Sensitization materials may include leaflets, booklets, posters, and videos.
56. Sensitization camps focus on promoting gender bias.
57. Gender sensitization encourages reinforcing traditional gender roles.
58. Gender sensitization is not relevant to the livestock sector.
59. Gender sensitization promotes shared responsibilities between men and women.
60. Gender sensitization is about challenging traditional gender roles and stereotypes.
61. Gender sensitization primarily encourages and maintains gender bias.
62. Gender sensitization can help individuals re-evaluate their attitudes and beliefs about gender.
63. Gender sensitization is only relevant to women in society.
64. Accommodation is one of the stages of gender sensitization.
65. Gender sensitization aims to maintain traditional gender roles and stereotypes.
66. Gender sensitization for veterinarians involves promoting empathy and understanding of the opposite gender's challenges.

67. The action stage of gender sensitization involves the implementation of policies and programs to improve women's conditions.
68. Gender sensitization primarily focuses on maintaining traditional gender roles and stereotypes.
69. The accommodation stage of gender sensitization involves breaking down barriers between men and women.
70. Affirmative action's aim to worsen women's conditions and reduce their participation in development.
71. Mass media and plays are ineffective ways to promote gender sensitivity.
72. Gender sensitization is essential for achieving gender equity in the livestock sector.

Question 3. Fill in the Blank

1. Gender sensitization involves the modification of behaviour through awareness of ____________ concerns.
2. Gender sensitization promotes empathy and understanding of ____________ roles and challenges.
3. In the livestock sector, male veterinarians can actively seek opportunities to work with ____________ pets to challenge stereotypes.
4. Change in perception, recognition, accommodation, and action are the four stages of ____________ sensitization.
5. Gender-sensitive modules may contain case studies and ____________ analysis in the livestock sector.
6. Gender sensitization encourages individuals to modify their behavior through awareness of ____________ equality concerns.
7. In the context of veterinary medicine, gender sensitization aims to encourage both male and female veterinarians to put themselves in the shoes of their colleagues of the ____________ gender.
8. In the stage of ____________ in gender sensitization, there is a change in perception about gender roles.
9. A key aspect of gender sensitization is breaking down ____________ in various fields, including veterinary medicine.
10. Gender sensitization involves addressing and challenging ____________ disparities in society.
11. Gender sensitization seeks to modify behaviour through ____________ of gender equality concerns.

12. The recognition stage of gender sensitization involves men coming around to recognize the virtues of women and their importance to the family and __________.
13. In the livestock sector, gender sensitization can encourage men and women to work together in all aspects of livestock farming, breaking down traditional gender __________.
14. Gender sensitization involves challenging traditional __________ and stereotypes.
15. In the livestock sector, gender sensitization is essential for achieving __________ and enhancing productivity.
16. Sensitization programs should target both men and women involved in __________ farming.
17. The "Action" stage in gender sensitization involves individuals acting as instruments of __________.
18. Gender-sensitive modules may contain case studies and situation __________.
19. Gender sensitization materials often include leaflets, booklets, posters, and __________.
20. Sensitization camps encourage participants to understand and discuss __________ issues.
21. Gender sensitization promotes shared responsibilities between __________ and __________.
22. The primary purpose of gender sensitization is to challenge traditional gender roles and promote __________.
23. Changing stereotypes and mindsets is one of the key goals of gender __________.
24. Gender sensitization is the modification of behaviour through awareness of __________ equality concerns.
25. A key component of gender sensitization is instilling __________ and understanding regarding gender issues.
26. In gender sensitization, men and women examine their personal __________ and beliefs.
27. Gender sensitization promotes breaking down barriers and fostering __________ between genders.
28. In the livestock sector, women are often associated with the care of __________ animals.

29. One of the stages of gender sensitization is a change in __________.
30. Gender sensitization aims to challenge traditional gender roles and __________.
31. Accommodation in gender sensitization refers to __________ between men and women.
32. Gender sensitization helps individuals see both genders as __________ partners in socio-economic development.
33. An example of behaviour modification in gender sensitization could involve men actively seeking opportunities to work with _______ pets.
34. Female livestock farmers should be encouraged to actively participate in __________ and decision-making.
35. Gender sensitization programs should focus on __________ and questioning stereotypes.
36. Changing stereotypes and traditional gender roles is a primary goal of __________ in animal husbandry.
37. Affirmative actions can be taken to improve women's conditions and promote gender __________.
38. Gender-sensitive programs can target women who are socially and economically __________.
39. Gender sensitization is essential for achieving gender equity and improving the productivity and __________ of livestock farming.
40. Gender sensitization often involves creating awareness about the challenges and issues that __________ face.
41. Sensitization programs can educate elderly women about the negative consequences of gender __________.
42. Sensitization programs can target __________ who initially resist changing gender roles due to cultural norms.
43. Sensitization can create a class of responsive functionaries at different levels who consider addressing gender-related issues as a matter of __________.
44. Gender sensitization can lead to increased sensitivity towards __________ and their problems.
45. Through gender sensitization, community members can understand that addressing gender bias is essential for achieving a more equitable __________ and economic order.

46. Sensitization programs should encourage men and women to work together in all aspects of __________ farming.
47. Gender sensitization highlights the importance of recognizing and addressing gender __________.
48. Gender sensitization encourages both men and women to participate in traditionally gendered tasks like __________.
49. Sensitization modules may contain __________ that highlight the roles of men and women in livestock management.
50. Gender sensitization often involves using materials like __________ to educate people about gender equity.
51. Community discussions and storytelling sessions can be used to educate elderly women about the ill effects of __________.
52. In gender sensitization, theatre plays can be organized to present real-life scenarios and challenge gender __________.
53. Gender sensitization emphasizes that men and women have to function in different socio-economic __________ within the livestock sector.
54. Sensitization programs can involve both men and women of various age groups who are engaged in ______ farming within the same household.
55. Gender sensitization in the livestock sector aims to make people more aware of the challenges and issues that __________ face.
56. Gender sensitization can help break down the stereotype that only men should work with __________ livestock.
57. Sensitization camps encourage open __________ and problem-solving related to gender biases in the sector.
58. Mass media campaigns in the livestock sector challenge stereotypes and showcase successful stories of gender __________.
59. Participatory discussions among men and women from different age groups help foster __________ and provide a platform for sharing practical experiences.
60. In gender sensitization, elderly women can learn about the positive outcomes of sharing livestock responsibilities and benefits from __________ gender roles.
61. Gender sensitization aims to break down barriers and foster __________ between genders in the livestock sector.
62. Gender sensitization focuses on spreading the message of how both men and women can contribute to family welfare, growth, and __________ of their villages.

63. Sensitization programs can target socially and economically progressive women who serve as __________ for underprivileged women.
64. Gender sensitization modules should be easily understandable by the __________.
65. Content in gender sensitization programs can include training sessions on animal __________ and farm management.
66. Sensitization materials like leaflets and posters should be designed to educate people about __________ equity in the livestock sector.
67. Sensitization programs aim to make people more aware of the benefits of involving women in __________ management.
68. Researchers can conduct studies to identify and develop better livestock breeding techniques suitable for small-scale __________ farmers.
69. Sensitization programs should initiate friendly debate among a larger audience on the ill effects of different forms of __________ bias.
70. Sensitization programs create awareness about the negative consequences of gender bias on women, men, families, and __________.
71. Gender-sensitive modules can contain case studies and situation analysis related to gender roles in the __________ sector.
72. Sensitization programs in the livestock sector aim to promote gender sensitivity, which is essential for achieving gender equity and enhancing the __________ and sustainability of livestock farming.

ANSWERS

1. Multiple Choice Questions

1	b	Challenge traditional roles
2	b	Empathy and understanding
3	c	Address gender disparities and promote equality
4	c	Modifying behavior through gender equality awareness
5	d	Recognition
6	b	Challenging Sterotypes
7	b	Modifying human behaviour through awarness of gender equality owens
8	c	Challenge and change their behaviour
9	d	Change in perception
10	c	Roles and contributor of both mens women

11	b	Challenge traditional gender roles and stereotypes
12	c	To promote gender equality and inclusivity
13	b	Gender sensitization
14	b	Both men and women involved in livestock farming
15	b	Organizing workshops for women to discuss their concerns
16	b	Case studies and situation analyses
17	b	Leaflets, booklets, posters, and videos
18	c	Understand and discuss gender issues
19	b	Changing stereotypes and mindsets
20	b	Shared responsibilities between men and women
21	c	To raise awareness of gender equality concerns
22	b	To encourage behavior modification
23	a	Change in perception
24	c	Challenging stereotypes
25	c	Target audience
26	c	To educate people about gender equity
27	b	Recognition
28	b	To challenge stereotypes and improve the sector's productivity
29	b	Challenge sterotypes and question gender biases
30	b	Promoting gender equality and awareness
31	d	Isolation
32	a	Traditional gender roles and stereotypes
33	b	Recognizing marks of women
34	b	Both mens women
35	c	Improved livestock management
36	a	Mass media compaigns
37	b	Taking steps to address gender bias & proriety equality
38	b	To achieve gender equality and impore livestock practes
39	c	Promoting shared responsibilities and challenging stereotypes
40	c	Tailoring advice to the specific needs of clients regardless of gender
41	b	It encourages individuals to work toward gender equality and create a favorable environment for women.
42	b	By showcasing successful stories of gender equality

43	c	To provide a platform for open dialogue and understanding of gender issues
44	b	Provide training and resources tailored to women's needs
45	c	Awareness
46	c	By actively working towards behavior modification to promote gender equality
47	c	Both men and women
48	c	Overcoming traditional barriers
49	b	By challenging stereotypes and sharing responsibilities
50	b	They can serve as role models and mentors to empower underprivileged women
51	d	Changing perceptions about gender
52	c	To promote gender equality and enhance productivity

Answer 2. True and False

S.No.	True or false	S.No.	True or false
1	True	2	True
3	False	4	True
5	True	6	True
7	False	8	True
9	False	10	True
11	False	12	False
13	True	14	True
15	False	16	False
17	True	18	True
19	False	20	True
21	True	22	True
23	False	24	True
25	False	26	False
27	False	28	True
29	False	30	True
31	False	32	True
33	True	34	False
35	False	36	False
37	True	38	False
39	False	40	True

41	True	42	False
43	False	44	False
45	False	46	True
47	False	48	False
49	False	50	False
51	True	52	False
53	True	54	False
55	True	56	False
57	False	58	False
59	True	60	True
61	False	62	True
63	False	64	True
65	False	66	True
67	True	68	False
69	True	70	False
71	False	72	True

Answer 3. Fill in the Blanks

1	Gender equality	2	Opposite sex's
3	Smaller	4	Gender
5	Situation	6	Gender
7	Opposite	8	Recognition
9	Stereotypes	10	Gender
11	Awareness	12	Society
13	Roles	14	Gender roles
15	Gender equality	16	Livestock
17	Change	18	Analyses
19	Videos	20	Gender
21	Men, women	22	Inclusivity
23	Sensitization	24	Gender
25	Empathy	26	Attitudes
27	Accommodation	28	Smaller
29	Perception	30	Stereotypes
31	Understanding	32	Equal
33	Smaller	34	Discussions
35	Education	36	Gender sensitization

37	Equality	38	Progressive
39	Sustainability	40	Women
41	Bias	42	Women
43	Priority	44	Women
45	Social	46	Livestock
47	Bias	48	Milking
49	Case studies	50	Posters
51	Gender bias	52	Stereotypes
53	Spaces	54	Livestock
55	Women	56	Large
57	Dialogue	58	Equality
59	Understanding	60	Changing
61	Accommodation	62	Development
63	Mentors	64	Audience
65	Husbandry	66	Gender
67	Livestock	68	Women
69	Gender	70	Society
71	Livestock	72	Productivity

4

The Importance of Gender Analysis in Livestock Farming

Chapter Overview

The need for gender analysis in various contexts, including the livestock sector, is vital for understanding and addressing differences and inequalities between men and women. This chapter explores the crucial role of gender analysis in the context of the livestock sector, focusing on animal husbandry and related activities. It delves into the significance of understanding and addressing gender disparities in livestock farming, highlighting the unique challenges and opportunities experienced by both men and women. The chapter emphasizes the need for equitable and sustainable development within the sector. It discusses how gender analysis can reveal disparities in roles, resource access, and decision-making power, and offers strategies to overcome challenges faced by women in livestock farming. The chapter underscores the potential benefits of promoting gender equality, such as increased productivity, economic empowerment, and overall well-being. The chapter also includes a section on the broader need for gender analysis as a tool to promote inclusivity and eliminate gender discrimination in various sectors and organizations. Overall, this chapter provides a comprehensive overview of why gender analysis is essential in the livestock sector, particularly in dairy farming, and how it can lead to more equitable and sustainable livestock production practices. The Chapter also mentions different frameworks for conducting gender analysis, such as the Harvard Analytical Framework, Moser's conceptual framework, the Social Relations approach, and the 4R method, which all provide structured methods for understanding and addressing gender disparities in various contexts, including the livestock sector.

SECTION A: THEORY

Concept of Gender Analysis

In various contexts, distinctions among women, men, girls, and boys become apparent in the distribution of resources, opportunities, restrictions, and power. These disparities are illuminated through a gender analysis, which delves into the specifics of how responsibilities are divided among them, how they manage these tasks, access resources, exert control over them, and navigate societal interactions. Moreover, it entails an examination of other aspects of gender-related norms, including those related to identity and sexuality. Gender analysis is essential for gathering the necessary information to promote gender inclusivity and involves exploring other gender expression norms, such as those related to sexuality and identity. Gender analysis forms the foundational data required for gender mainstreaming.

Ester Boserup (1970) was first to challenge the notion that development would automatically improve women's status. In recognizing that women's position relative to men could worsen because of development as currently practiced, Boserup's work launched the field of research on women in development (WIl)) around the theme of equity.

Poats and Feldstein (1989) stated that "By recognizing differences between men's and women's roles in production, the assumed homogeneity of the farm household was replaced by the concept of "intrahousehold dynamics." The recognition that these diverse and complex relationships among members of households must be considered in the design, testing and evaluation of technology has stimulated some of the most exciting and innovative methodological developments in FSR/E to date."

Gender analysis begins with the recognition that the household is not an undifferentiated grouping of people with a common production and consumption function, that is, with shared and equal access to resources for and benefits from production. Rather, households are themselves systems of resources allocation (Guyer 1980). The pattern of decision making varies from one place or culture to another. Gender analysis focuses on differences in the activities, resources, and benefits of different members within the household and on patterns of obligation, cooperation or conflict between household members.

In various contexts, distinctions among women, men, girls, and boys become apparent in the distribution of resources, opportunities, restrictions, and power. These disparities are illuminated through a gender analysis, which delves into the specifics of how responsibilities are divided among them, how they manage these tasks, access resources, exert control over them, and navigate societal interactions. Moreover, it entails an examination of other aspects of gender-related norms, including those related to identity and sexuality. Gender analysis is essential for gathering the necessary information to promote gender inclusivity and involves exploring other gender expression norms, such as those related to sexuality and identity. Gender analysis forms the foundational data required for gender mainstreaming.

The definition of gender analysis can vary in diverse contexts but essentially involves the systematic gathering and evaluation of both quantitative data (numerical information) and qualitative insights (such as preferences, beliefs, attitudes, behaviours, and values) while taking gender-related aspects into account. This analytical approach aims to scrutinize the differences in societal roles and norms between different genders, such as men and women, boys and girls, and how these disparities influence their lives.

Unfortunately, some think that 'gender analysis' represents a kind of 'gender warfare' that challenges the dominance of males in the household and the society at large. This general attitude hinders the success of many development programs, because even the choice of the researcher for gender analysis is not gender neutral. Both male and female researchers can be equally proficient at gender analysis, and both can be equally 'gender blind'.

Gender blindness is the inability to perceive different gender roles and responsibilities. Gender blindness implies, for example, the perception that all farmers are males, and that research and extension activities have the same effect on men and women. For a deeper understanding of the gender issues in rural development it is important to consider the following factors:

- The division of labour,
- Access to and control over productive resources,
- Stakes and incentives in project activities,
- Contribution to household income,
- Degree of income pooling within the household,
- Responsibilities for different types of expenditure.

Gender analysis examines and highlights the dynamics among individuals of different genders within society and the imbalances in those dynamics. It seeks to answer questions like: Who performs various roles? Who has ownership of different resources? Who holds authority and how? Who experiences advantages? Who faces disadvantages? While asking these questions, it also considers the specific individuals within these gender categories. Gender analysis connects personal relationships within the private sphere with broader societal connections in the public domain. Gender roles and expectations learned within families are often reinforced and replicated in the public domain. For example, if a girl grows up seeing her mother primarily responsible for domestic tasks while her father assumes the role of the breadwinner, she might internalize these roles and replicate them in her own adult life. It also explores how gender norms intersect with global economic systems, national policies, cultural norms, and local practices. For instance, the gender wage gap at a global level might be connected to workplace policies and societal expectations at a national level, which in turn are influenced by family dynamics and cultural norms at a local level.

Definition of Gender Analysis

Gender Analysis is defined as analysis focused on the relative distribution across genders of "resources,opportunities, constraints and power in each context."

(SIDA 2015)

Gender analysis is 'the study of differences in the conditions, needs, participation rates, access to resources and development, control of assets, decision-making powers, etc., between women and men and their assigned gender roles.'

(European Commission, 1998)

Gender analysis is an analytical, social science tool that is used to identify, understand, and explain gaps between males and females that exist in households, communities, and countries, and the relevance of gender norms and power relations in a specific context.

(USAID, 2013)

Gender analysis refers to the systematic gathering and analysis of information on gender differences and social relations to identify and understand the different roles, division of labour, resources, constraints, needs, opportunities, and interests of various groups, including men and women, girls and boys, and transgendered person in each context. It aims to clarify how gender roles and relations create opportunities for, or obstacles to, achieving development objectives.

(Manfre *et al.*, 2013)

Such analysis typically involves examining differences in the status of women and men and their differential access to assets, resources, opportunities, and services. It considers the influence of gender roles and norms on the division of time between paid employment, unpaid work (including subsistence production and care for family members), and volunteer activities. It also examines the impact of gender roles and norms on leadership roles and decision-making, constraints, opportunities, and entry points for narrowing gender gaps and empowering females, and potential differential impacts of development policies and programs on males and females, including unintended or negative consequences.

Gender mainstreaming and Gender Analysis

Gender mainstreaming is a worldwide strategy to promote fairness between genders in all parts of society. It was agreed upon at the Fourth United

Nations World Conference on Women in Beijing in 1995. Since then, many governments in both developing and developed countries, have started using this strategy.

Gender mainstreaming means making sure that gender equality is a part of everything we do in society. It involves considering the perspectives, experiences, knowledge, and interests of people of all genders in every aspect of society. This includes politics, economics, the environment, social issues, culture, and institutions.

The United Nations recognizes gender analysis as a crucial first step for gender mainstreaming. Gender analysis and other programs that raise awareness about gender have been widely used as tools to put this strategy into action and achieve fairness between genders. Gender mainstreaming remains a strategy, gender analysis may be a starting point along with awareness programs and goal is gender equality.

Critical Aspects of Gender Analysis

A comprehensive gender analysis should encompass the following critical aspects:

1. Political and Legal Framework

a) Gender-Responsive National Policies and Laws: Examining whether national policies, laws, and regulations acknowledge gender equality and women's rights. It involves assessing the extent to which these legal frameworks promote gender equity.

b) International Commitments: Evaluating whether national policies and regulations align with international commitments on gender equality, such as standards set by the Council of Europe, the Convention on the Elimination of All Forms of Discrimination against Women (CEDAW), and the United Nations Sustainable Development Goals (SDGs).

For example, in an animal husbandry sector of a country, there are national policies and laws related to livestock farming. Through gender analysis, it should assess whether these policies adequately consider gender equality and women's rights in the livestock sector and whether women have sufficient access to government support programs and resources. The analysis should further indicate that these policies are aligned with the country's international commitments to gender equality, such as the SDGs.

2. Access to and Control over Resources

i. Resource Definition: Recognizing that resources encompass not only material and financial assets but also intangible resources such as time, information, knowledge, and rights.

ii. Availability and Access to Resources: The resource like cash is available in the house but the women may not have access to these resources. Investigating who has access to these resources and who benefits from them is an important issue which has a bearing on their participation in activities. This involves understanding the distribution of resources among different gender groups.

iii. Control over Resources: Analysing who can obtain access to resources and make decisions about their utilization. This includes identifying factors that influence control over resources.

iv. Awareness of Legal Rights: Assessing whether women and men are aware of their legal rights regarding access to and control over resources. This is crucial to determine if individuals can assert their rights effectively.

For example, gender analysis in animal husbandry should highlight whether the access of men and women to financial and livestock-resources is equal. Unequal access to and control over resources between men and women may affect women's ability to participate effectively in animal husbandry programmes.

3. Access to Services and Institutions

i. Availability and Equality: Evaluating the availability of various services and institutions for women and men and assessing whether there is equal access for both genders need to be assessed through gender analysis.

ii. Treatment by Service Providers: Examining whether women and men are treated equally by service providers and whether there are any biases or discrimination in service provision also need to be probed.

iii. Affordability: Analysing whether services and institutions are affordable for both women and men, considering their socio-economic circumstances.

For example, gender analysis might reveal that veterinary services in a rural area are primarily available during working hours, making it challenging for women, who often have domestic responsibilities, to access these services during this period. Additionally, women are not treated equally by service providers especially the male and may face gender biases in terms of the quality of service they receive. Services are not affordable for all community members, leading to disparities in accessing animal healthcare services.

4. Women's and Men's Roles

Productive Work: Investigating gender roles in income-generating work, including whether there is horizontal segregation (women and men in certain expected professions) or vertical segregation (different gender groups occupying positions of varying authority in the labour market).

Household Work: Analysing the division of household tasks, particularly unpaid care work, which often falls on women. This includes food preparation, childcare, and maintaining the home.

Community Work: Assessing the involvement of women and men in community activities and their respective roles in political, religious, or social organizations. This also involves understanding whether participation is voluntary or incentivized with power, status, or monetary rewards.

5. Participation in Decision-Making

Who Participates: Identifying the individuals who participate in decision-making processes and at which levels (household, community, or organizational).

Decision-Making Procedures: Analysing the methods used for decision-making and whether they consider the needs and perspectives of women and men equally.

For example, gender analysis may reveal that decision-making in the community's livestock management is largely male dominated. Men participate in discussions and decisions related to breeding and selling of animals. Women have limited influence in these processes. The decision-making procedure does not adequately consider the needs and perspectives of both women and men, leading to disparities in livestock management practices.

A robust gender analysis, based on these key aspects, provides valuable insights into the dynamics of gender roles, rights, and disparities. It informs the development of policies, programs, and projects that are more equitable, responsive, and inclusive of both women and men, contributing to the advancement of gender equality and women's empowerment. For instance, programs can be designed to provide women with access to veterinary services outside regular working hours, training opportunities in animal healthcare at a venue and time suitable for women to participate, and more equitable participation in decision-making processes. These actions aim to address gender disparities and promote gender equality within the livestock sector.

Steps to conduct a gender analysis

Steps to conduct a gender analysis

The following are three suggested steps to take when carrying out gender analysis:

a) Collection of Gender- and Sex-Disaggregated Data:

Collection of Gender- and sex-disaggregated data and information, quantitative and qualitative to enable the project team to see the need for gender mainstreaming activities. Possible sources of information can be official statistics, research reports and reports published by various organizations including NGOs. Ideally, information should be collected also by consulting stakeholders and target groups of the project. It is also worth gathering lessons learned from other organizations and similar interventions. Gender-disaggregated data is required for setting the baseline of indicators. It is important to keep in mind that neither women nor men are a homogenous group, and that intersectionality occurs, which means the overlap of gender with age, disability, minorities, sexual orientation, social identities, socio-economic status, etc. For example, in an animal husbandry project, the project team recognizes the need to collect gender- and sex-disaggregated data to understand how men and women are involved in livestock management. They collect quantitative data, such as time spent on animal care by men and women and the income generated from livestock activities. Qualitative data is also gathered through focus group discussions with women and men in the community.

They consult with local stakeholders, including female and male livestock keepers, to gain insights into their roles and challenges. Additionally, they review published research reports and case studies which provide valuable information on gender dynamics in livestock farming. The data collected may reveal that women are primarily responsible for small ruminants and Backyard poultry, while men are more involved in herding larger animals. This information highlights the gender disparities in livestock management.

b) Identifying gender differences and the underlying causes of gender inequalities:

A good gender analysis will show the needs of these groups and will thus allow tailoring the project or programming to the existing needs in society. For example, in animal husbandry, through gender analysis, the project team identifies that women's limited access to veterinary services is a significant gender difference and a cause of gender inequality in animal husbandry. Men often make decisions about when to seek veterinary care for livestock, which can result in delayed or inadequate healthcare for animals. This gender difference leads to increased livestock morbidity and mortality among female-headed households.

By understanding this difference and its underlying causes, the project may alter its activities and tailor interventions to address the specific needs of women in accessing veterinary services. For instance, they may establish mobile veterinary clinics that visit communities at times convenient for women, thus reducing the gender gap in access to animal healthcare.

c) Informing the development policies, programs, and projects that respond to the different needs of women and men:

A rigorous gender analysis will ensure that sound and credible advice is provided, and the policies, programs, and projects developed based on it will have greater credibility and validity among those affected by them. For example, building on the gender analysis findings, the project team advises local government authorities and agricultural development organizations on the need to address gender disparities in animal husbandry. They present evidence of gender-based differences in livestock care and unequal access to veterinary services.

In response to this advice, the local government implements a policy that ensures equal access to veterinary services for both men and women livestock keepers. Furthermore, they design a training program specifically tailored to the needs of women in the community. This program equips women with the knowledge and skills to take on more active roles in livestock healthcare decision-making, ultimately leading to improved animal health and increased household income.

By following these steps, the animal husbandry project becomes more gender-responsive and effective, addressing the specific needs and challenges faced by women and men in livestock management and promoting greater gender equality in the sector.

Various types of Gender Analysis Frameworks

1. Harvard Analytical Framework

The Harvard Analytical Framework for gender analysis was developed in 1985 with an aim to demonstrate an economic case for allocating resources for women as well as men at a time when the efficiency approach was gaining prominence (March, Smyth and Mukhopadhyay 1999). The framework has four main components – three tools for gender analysis and a checklist to examine a project proposal or intervention from a gender perspective using gender-disaggregated data and capturing the different effects of social change on men and women. The three tools for gender analysis are discussed in brief herewith:

Tool 1: The Activity Profile

This tool identifies all relevant productive and reproductive tasks and answers the question: who does what? How much detail you need depends on the nature of your project. It is advisable to add a time dimension – specifying what percentage of time is allocated to each activity, whether it is carried out seasonally or daily; or a skill and technology dimension – specifying whether the activity involves only manual labour, or specific skills and tools for undertaking. For instance, in the realm of animal husbandry, it helps identify the individuals responsible for tasks like feeding, cleaning, milking, and administering medical care to the animals. While such tasks may be shared between men and women in many societies, there may be gender-specific divisions, with men often handling physically demanding activities and women taking care of tasks requiring gentler care.

Figure: An example of activity profile for dairy activity in south Asian context

S.No.	Dairy Activity	Majority of the work done by	
		Male	Female
1	Arranging loans for buying of animals		
2	Buying of animals		
3	Cleaning		
4	Bathing		
5	Feeding		
6	Milking		
7	Taking cattle for grazing		
8	Getting fodder from field		
9	Buying fodder		
10	Selling of milk		
11	Collecting money from selling milk		
12	Insurance of animals		
13	Availing veterinary services		
14	Selling of animals		

Tool 2: Access and Control Profile (Resources and Benefits)

This tool enables users to list what resources people use to carry out the tasks identified in the Activity Profile. It indicates whether women or men have access to resources, who controls their use and who controls. Source: Adapted from March, Smyth and Mukhopadhyay (1999). the benefits of a household's (or a community's) use of resources. Access simply means that you can use a resource. The person who controls a resource is the one ultimately able to make decisions about its use, including whether it can be sold. In the context of animal husbandry, this involves identifying resources like grazing land, animal feed, veterinary supplies, and equipment. Gender roles can play a significant role in determining access and control, with men often making decisions regarding land use and significant expenses, while women may oversee day-to-day expenses and tasks.

Figure: A sample for dairy activity in south Asia context (Access and control)

Tools/Resources/Decisions required	Access		Control/Ownership	
Tools and Resources	**Male**	**Female**	**Male**	**Female**
Credit for Cattle				
Cattle				
Cattle shed				
Feed services				
Veterinary services				
Insemination services				
Fodder availability				
Grazing lands/common plots				
Extension services				
Milk cooperative membership				
Milk cooperative position holder				
Milk cooperative union(district-level) membership				
Decisions required	**Male**	**Female**	**Male**	**Female**
Which cattle breed to purchase				
Number of cattle to keep				
Disposal of non-milch cattle				
Maintenance of grazing land				
Fodder production in own field				
When to call the veterinarian vs local treatment				
Selling of calves/cattle				
How much milk to sell vs how much to keep for home consumption				

Tool 3: Influencing Factors

This tool helps chart factors which influence the differences in the gender division of labour, access and control as listed in the two Profiles (Tools 1 and 2). Influencing factors include all those that shape gender relations and determine different opportunities and constraints for men and women. These factors are far-reaching, broad and interrelated. This tool is intended to help you identify external constraints and opportunities which you should consider in planning your development interventions. Factors such as cultural norms and expectations may dictate that men hold decision-making authority in significant transactions, while women are responsible for tasks like milking and dairy product preparation. Economic factors, including income distribution within the household, can also play a pivotal role in determining decision-making power.

The Harvard Analytical Framework serves as a valuable tool for comprehending gender dynamics within various activities, resource allocation, and control over benefits. It consists of three interconnected components, each shedding light on distinct aspects of gender roles and their influence. It's crucial to recognize that while the Harvard Analytical Framework provides valuable insights into gendered roles and resource distribution, it does not inherently challenge existing gender inequalities. In the context of animal husbandry, the framework aids in understanding the dynamics at play but does not directly confront traditional gender norms. Challenging and transforming gender inequalities necessitates additional strategies and interventions that go beyond the framework itself. The framework primarily serves as a tool for analysis and comprehension, requiring complementary efforts to explicitly address and rectify gender inequalities.

2. Moser conceptual framework

The Moser Framework, pioneered by Caroline Moser during the early 1980s, strives to establish gender planning as an independent process. This framework comprises six tools rooted in three core concepts:

i. Identifying gender roles and the triple burden of women's work (productive, reproductive, and community).

ii. Addressing practical needs and strategic gender interests.

iii. Recognizing the approaches of Women in Development (WID) and Gender and Development (GAD) policies.

Tool 1: Identifying Gender Roles and Women's Triple Burden of Work:

This tool aids in recognizing women's triple burden of work:

A. Reproductive roles encompass caring for the household and its members, including child-rearing, food preparation, water and fuel collection, laundry, shopping, housekeeping, and family health care.

B. Productive roles involve producing goods and services for consumption and trade, whether in formal or informal employment or self-employment.

C. Community work entails organizing social events and services, community improvement activities, group participation, local politics, and more.

While both genders may participate in these areas of work, men are typically less involved in reproductive tasks. Moreover, the roles and responsibilities in productive activities often differ between men and women, with women's contributions often undervalued. Women are often engaged in managing community resources like water, education, and healthcare, while men are more involved in community politics and formal decision-making processes.

Tool 2: Gender Needs Assessment

The idea of women's practical and strategic interests was originally developed in the 1980s by Maxine Molyneux, and later by Caroline Moser.

A. Practical gender needs are those which, if they were met, would assist women in their current activities without challenging the existing gender division of labour. These include:

i) Water provision; ii) Healthcare provision; iii) Opportunities for earning an income to provide for the household; iv) Provision of housing and basic services; vi) Distribution of food; and others.

These needs are shared by all household members, yet women often identify them as their specific needs because it is women who assume responsibility for meeting their families' requirements. For instance, in the livestock sector, a practical gender need might be related to women's daily tasks of milking and feeding livestock. Women may need access to training in modern milking techniques or improved access to water sources for livestock

B. Strategic gender interests are those which exist because of women's subordinate social status. If met, these would enable women to transform existing imbalances of power between women and men. These relate to gender divisions of labour, power and control, and may include such issues as legal rights, domestic violence, equal wages and women's control over their own bodies. Strategic Gender Needs are the needs that have the potential to transform existing gender subordination. In the livestock sector, a strategic gender need could involve challenging the traditional gender roles where men have control over major decisions related to animal breeding and healthcare.

Tool 3: Disaggregated control of resources and decision-making within the household

This tool links allocation of resources within the household (intra-household allocation) with the bargaining processes which determine this. Who has control over what resources within the household, and who has what power of decision-making?

Tool 4: Planning for balancing triple roles

This tool looks at the impact of a project intervention on women's triple work burden. Sectoral planning frameworks, which concentrate only on one role, often tend to ignore the effect on women's other roles. Users of the framework are asked to examine, whether a planned programme or a project will increase a woman's workload in one of her roles, to the detriment of her other roles. For example, the provision of irrigation water will increase women's participation in agriculture activities while constraining the time available for domestic activities or might increase the workload of fetching water due to diversion of fresh water from domestic use to irrigation.

Tool 5: Distinguishing between different aims in intervention

This tool helps identify the approach that a project followed or will follow (if used for evaluation) by asking to what extent do different approaches meet practical and/or strategic gender needs.

Moser classified various policy approaches into five categories based on this:

1. Welfare approach which focuses on practical gender needs and sees women as passive beneficiaries of development interventions;
2. Equity approach which focuses on strategic gender interests and recognizes women as active participants in development;
3. Anti-poverty approach which focuses on practical gender needs and ensures that poor women move out of poverty by focusing on increasing their productivity;
4. Efficiency approach which recognizes all three roles but focuses on practical gender needs for harnessing women's economic contribution; and
5. Empowerment approach which focuses on strategic gender interests through supporting their own initiatives, thus fostering self-reliance. This approach recognizes women's subordination not only because of male oppression but also because of colonial and neo-colonial oppression.

Tool 6: Involving women and gender-aware organizations and planners in planning

Finally, Moser's framework asks users to think about the importance of involving women, gender-aware organizations and planners themselves in planning. This should be at all levels – in planning, implementation and monitoring and evaluation.

3. The Social Relations Framework (SRF) Approach

The Social Relations framework approach, developed by Naila Kabeer (1994) in collaboration with policymakers, academics, and activists at the Institute of

Development Studies, Sussex University, UK, is based on several theoretical foundations. It differs from the other frameworks in that it engages in an analysis of institutional relations rather than focusing on roles, resources, and activities (as in the Harvard Framework and the Gender Analysis Matrix). It takes the four key institutions of society to be the State, the market, the community, and the family or kinship. It encourages analysis between these institutions as well as within a single institution This approach is intended as a method of analysing the gender inequalities within institutionalised relations that affect the distribution of resources, responsibilities, and power. The Social Relations Approach can be used in a narrow application to analyse how gender inequality is formed and reproduced within a single institution. Alternatively, it can be applied more broadly to reveal how gender and other inequalities are interlinked through interaction between different institutions, creating situations which disadvantage certain individuals and groups in multiple way

Five aspects of social relations shared by institutions

Although institutions vary in their purpose, their culture, and their working practices, according to Kabeer they share five distinct but inter-related components of social relationships: rules, resources, people, activities, and power. Analysing institutions according to these categories helps us to understand why some people gain and others lose out

1. Rules: how things get done Institutional behaviour is governed by rules. These may be official and written down or unofficial and expressed through norms, values, traditions, and customs
2. Resources: what is used, what is produced? Institutions mobilise and distribute resources; these may be human resources (labour, education, skills, etc.), material resources (food, assets, land, money), or intangible ones (information, political power, goodwill, or contacts). As the Harvard Framework analysis shows, organisations often distribute resources according to social categories (gender, ethnicity, religion, for example) tied to rules.
3. People: who is in, who is out, who does what? Institutions are selective about whom they allow in and whom they exclude, who has access to various resources and responsibilities, and who is positioned where in the hierarchy. This selection may reflect class, gender, or other social inequalities
4. Activities: what is done? Institutions do things: they try to achieve things by following their own rules and ensuring routinised practice for carrying out tasks. These activities can be productive, distributive, or regulative. They consider who does what, and what they get for doing it

5. Power: who decides and whose interests are served? Institutions embody relations of authority and control. Few institutions are entire egalitarian even if they profess to be so. The unequal distribution of rules, resources, and activities ensures that some institutional actors have authority and control over others.

The gender analysis matrix framework promotes a bottom-up approach through community participation to examine how gender differences impact four crucial areas: labour, time, resources, and sociocultural factors. This framework offers a community-based method for identifying and analysing gender disparities at the societal, household, and individual levels, aiding communities in challenging and revaluating their assumptions about gender roles in a constructive manner. For example, applying the Social Relations approach to small-scale poultry farming, one would analyse how these dynamics influence gender relations and inequalities. Researchers might examine how a gender-sensitive policy impacts women's access to training and resources, leading to a more equitable division of labour and decision-making within poultry farming households. They may also assess whether increased participation of women in poultry farming enhances their economic well-being and status within the community.

Understanding the various elements of the Social Relations approach in this context can lead to more informed policies and interventions that aim to reduce gender inequalities and promote the well-being of both men and women involved in small-scale poultry farming.

Gender Policies

Gender-Blind Policies: Gender-blind policies consider men and women to be unequal and often reinforce gender norms, roles, and stereotypes that perpetuate gender inequalities. Imagine a government program that provides subsidies for poultry feed but does not consider the different roles of men and women in poultry farming. This can perpetuate gender inequalities if the subsidies are predominantly accessed by men, reinforcing traditional gender norms.

Gender-Neutral Policies: Gender-neutral policies operate within the existing gender division of resources and responsibilities without challenging it. These policies assume that men and women are not equal but are the same, leaving gender norms, roles, and relations unchanged. A gender-neutral policy might provide equal access to resources without recognizing the different roles of men and women. In this case, the policy doesn't challenge existing gender norms, and the workload and decision-making dynamics may remain unchanged.

Gender-Sensitive Policies: Gender-sensitive policies presume men and women as equals and address gender norms, roles, and access to resources to achieve policy objectives. If a policy acknowledges and addresses the different roles of men and women in poultry farming, it might provide training and resources to women to improve their poultry management skills. This can lead to more equitable gender relations within the sector.

Gender-Positive Policies: Gender-positive policies also presume men and women as equals and place a strong emphasis on changing gender norms, roles, and access to resources as a core component of policy outcomes. A gender-positive policy could go a step further by actively promoting women's participation in poultry farming, providing leadership opportunities, and challenging traditional gender norms. This can lead to greater empowerment and economic independence for women in the sector.

4. Gender Analysis Matrix

The Gender Analysis Matrix (GAM) was developed by Parker in 1993 to find out the different impacts of development interventions on women and men by providing a community-based technique for the identification and analysis of gender differences (March, Smyth and Mukhopadhyay 1999; Parker 1993). It also assists the community to identify and challenge their assumptions about gender roles in a constructive manner. The analysis is conducted at four levels of society: **women, men, household and community.** The GAM examines impact on four areas: labour, time, resources and socio-cultural factors. The GAM features these two main concepts on a matrix which focuses on the impact of the proposed development intervention.

	Women	**Men**	**Household**	**Community**
Labour				
Time				
Resources				
Culture				

Source: Parker (1993).

GAM TOOL 1: Analysis at Four 'Levels' of Society

GAM allows analysis of an intervention at four levels: men, women, households and community. The levels of analysis appear vertically on the matrix:

Men: Represent men of all ages who are in the target group or all men in the community.

Women: Represent women of all ages who are in the target group or all women in the community.

Household: Represents all women, men and children living under one roof (or extended family) as defined within the culture.

Community: Represents everyone in the community.

It is also important for the facilitator to account for age group, class, ethnic composition, social system (caste) and other important variables in the community.

GAM Tool 2: Impact Analysis

GAM examines impact on four areas, which appear horizontally on the matrix: Labour: Captures changes in tasks (Do women take over men's tasks in the field?), the level of skill (formal education, training) required, the number of people involved in this activity and the demand for additional labour.

Labour: This refers to changes in tasks (for example, fetching water from the river), the level of skill required (skilled or unskilled, formal education, training), and labour capacity (How many people carry out a task, and how much can they do? Is it necessary to hire labour or can members of the household do the work)?

Time: This refers to changes in the amount of time (three hours, four days, and so on) it takes to carry out the task associated with the project or activity.

Resources: This category refers to the changes in access to resources (income, land, and credit) because of the project, and the extent of control over changes in resources (more or less) for each group analysed.

Socio-cultural factors: This refers to changes in social aspects of the participants' lives (including changes in gender roles or status) because of the project.

Gender Analysis across Animal Husbandry Project Lifecycles: Design, Implementation, and Evaluation

Gender analysis is crucial throughout the life of a development project, starting from planning through implementation to assessment. During the planning stage, those responsible for economic and social project analysis should conduct gender analysis. The project management team can handle gender analysis during implementation, while at evaluation stages, input from a social scientist is beneficial.

For example, the BIOCON project in India initially focused on treating straw for mixed farming involving both men and women in crop and livestock activities. As the project progressed, it became evident that technology should be considered in the context of the users' living conditions. In a later phase, gender became a relevant aspect of the project, recognizing that in many parts of India, women are primarily responsible for caring for animals. They handle tasks like feeding, bathing, milking, cleaning sheds, and processing milk.

Despite often doing the work, men may receive the money. Including gender considerations later in the project increased awareness and led to the collection of gender-segregated data.

For instance, when it comes to high-yielding crop varieties, the amount of grain is more crucial for male farmers as it directly impacts their income. However, this may result in women having less straw, requiring additional effort to collect supplementary feed like greens to feed their livestock. This, in turn, increases the workload for women. Additionally, the perception of treated straw can vary for men and women, as it affects the quality of dung used for fuel cakes. In some cases, male farmers refused to treat straw if their wives could earn extra income without sharing it with them, highlighting the complex dynamics of gender roles in agriculture.

Conceptual Framework for Gender Analysis in Farming Systems Research and Extension (FSR/E) activities

Farming Systems Research and Extension (FSR/E) is an approach used in agricultural research and extension to develop suitable technology for specific clients, typically low-input or resource-poor farmers. FSR/E is holistic and iterative, emphasizing interdisciplinary teamwork to address entire farming systems rather than isolated crops or livestock. It views farms as interconnected systems, considering the potential impacts of new technologies on the entire farming system to avoid creating narrow, short-term solutions that may lead to unforeseen problems. The conceptual Framework for Gender Analysis in Farming Systems Research and Extension aims to correct the existing gender bias in agricultural research and extension. Currently, most projects consider the household as the primary unit, focusing on male heads of households as the key decision-makers. Unfortunately, this approach often overlooks the valuable contributions of other household members, such as adult women, senior men and women, and children. This oversight is not beneficial for the project and the people it intends to help. Women, seniors, and children bring unique skills, resources, and priorities to farming, and neglecting their involvement means neglecting a significant portion of the decision-making system in agriculture. This conceptual framework provides a guide for collecting and analyzing information about gender roles and household dynamics within farming systems. The goal is to use this information to design improved agricultural and livestock technologies that enhance yields, consumption, risk reduction, environmental stability, and overall farm productivity. The framework emphasizes the need to understand how male and female farmers are involved in research and extension activities in each area. In agricultural projects using the farming systems research and extension (FSR/E) approach, various types of data, including agroclimatic, biological, and socioeconomic information, are crucial for identifying problems and exploring technical possibilities for new technologies. Socioeconomic data, incorporated from the beginning, helps scientists comprehend farmers' decision-making processes—how they mobilize resources and their interest in specific agricultural enterprises, especially when introducing new or modified practices. Combining technical and social science data is essential for pinpointing research areas that align with farmers' needs.

In most societies, decision-making in farming is significantly influenced by gender roles and relationships within and between households. These dynamics are shaped by factors like gender differences, age, household position, class, ethnicity, and life cycle stages. Understanding these dynamics is essential for developing technologies that effectively address the diverse needs and circumstances of farmers.

Feldstein and Poats (1989) gave a comprehensive overview of gender analysis in Farming Systems Research and Extension (FSR/E) activities as discussed below:

1. Activities Analysis

In this part, we're looking at who does which tasks, especially considering the different seasons in farming (Table 1 a).

i) List all the tasks involved in farming, including growing crops, taking care of animals, managing household chores, and other activities that help the family or individuals, either by providing cash or goods able (1a).

ii) Once you have the tasks listed, you can mark who does each task using symbols or colours in the Activities Analysis table (1b). This helps identify gender roles, age groups, or other factors involved in doing specific tasks.

iii) Sometimes, certain activities are mainly done by one gender. For example, men might primarily handle cattle, while women focus on crops.

iv) In other cases, tasks within the same farming activity might be divided by gender. For instance, men might handle land preparation while women take care of weeding.

v) The seasonal calendar shows when labour is needed the most and helps identify all the tasks that compete for attention, categorized by gender, including tasks outside of farming.

vi) The Activities Analysis table indicates who does which task based on gender, age, and other factors. By understanding this, we can see whose work might be affected by changes, what tasks compete for their time, and who needs training in new methods.

Use of the two worksheets, separately or together, create activities map or profile with which to screen the identification of problems, the selection of research priorities, the designation of collaborating farmers, and the design of on-farm trials.

In cases where the use of labour or the timing of operations are not affected by proposed changes, activities analysis without the calendar may be sufficient.

Table 1a. Farming Systems Calendar. This worksheet serves as a visual representation of the farming systems calendar, allowing for a comprehensive overview of activities throughout the months and seasons.

Activity	Crop Production	Livestock	Household Production use	Off-farm activities
January				
February				
March				
April				
May				
June				
July				
August				
September				
October				
November				
December				

Table 1 b Activities Analysis. This worksheet is designed for a detailed analysis of various activities, differentiating between males and females. It covers crop production, livestock, household production, and off-farm activities.

Activities	Males	Females
Crop Production		
Crop/Field 1		
Task 1		
Task 2		
Task 3		
Crop/Field 2		

Task 1		
Task 2		
Task 3		
Crop/Field 3		
Livestock production		
Animal 1		
Task 1		
Task 2		
Task 3		
Household Production		
Off-farm Production		

2. Resources Analysis

Farm management decisions are influenced or determined by the availability of and control of or access to resources or inputs.

i) Resources Analysis helps break down who, by gender and age, can access and control important resources.

ii) When we talk about control, we mean having the authority to decide how a resource is used and allocated. Access refers to the freedom or permission to use a resource, sometimes with some decision-making power once access is granted.

iii) Examples help clarify this. For instance, if men control livestock for ploughing, their wives and female relatives may still use the animals for ploughing, but the men ultimately decide. Women may control cash and decide how it's spent, while men have access to it.

iv) Resources encompass various elements such as land (and its terms of availability), capital (including cash, tools, and livestock), labour (one's own, family, or hired), other inputs like seeds, fertilizers, and pesticides, as well as services like credit and education. It's essential to consider who has the means to purchase inputs if they're bought, and who participates in local exchanges for inputs. Regarding on-farm inputs like manure or mulch, it's important to recognize who controls or accesses them.

v) Knowledge is a crucial but often overlooked resource, which includes insights from years of farming experience, practical understanding of soil conditions, traditional risk-reducing techniques, and knowledge of seed varieties. Such knowledge is usually associated with specific tasks and is vital for understanding current resource usage and evaluating proposed changes.

vi) Additionally, resources include access to markets, which can be influenced by factors like mobility and membership regulations. While a distinction is made between labour and non-labour resources, it's important to identify instances where the use of one resource provides access to another, such as exchanging labour for land or access to traction animals.

vii) Access, and control analysis creates another screen or map for looking at production constraints and proposed solutions. What are the available resources? What resources are required for proposed changes? Who controls them & to whom and how will new resources be made available? Or whose resource shortage is relieved? Does the control or non-control of key resources suggest separate research domains?

Table 2: Resources Analysis

Resources	**Access**	**Control**	**Uses/Characteristics**	**Implication for FSR/E**
Land				
Water				
Labour				
Livestock				
Inputs Purchased Produced on Farm				
Cash				
Agricultural Credit				
Knowledge				
Markets/ Transportation				
Education				

3. Benefits and Incentive analysis

Benefit analysis looks at who benefits from or controls the output of production. This includes everything the product is used for, for example a crop such as rice for home consumption or selling it for money, making compost, fodder, building materials etc. It also considers other things that could be produced instead, like wild regions that might have valuable plants or animals, or other activities that compete for resources and time, like other types of farming or work.

Incentive analysis goes a step further by looking at why farmers choose to keep doing what they do or make changes. Farmers have preferences that influence their decisions. These preferences can be related to how well a crop grows, how much it yields or earns, if it helps protect the environment, reduces risks, fits their schedule, or requires less work. Preferences can also be about what the product is used for, like if it brings prestige, fulfils obligations to family or community, tastes good, sells well, improves nutrition, is easy to process, or provides fuel, animal feed, or building materials.

Benefits and Incentives Analysis," helps create a detailed breakdown, focusing on gender and age, of the reasons why people benefit from or are motivated to engage in agricultural or other types of production. This breakdown helps identify what qualities should be added or kept in new plant varieties or technologies. It also highlights what might be lost in terms of characteristics and suggests possible alternatives. The goal is to assess whether proposed changes in activities or resource use align with who benefits from them. This analysis helps determine if those who bear the extra costs are the ones who gain the most from the changes.

Table 3: Benefits and Incentive Analysis

Stages of FSR	**Access**	**Control**	**Uses/Characteristics**	**Implication for FSR/E**
Crop Production				
Livestock production				
Household production				
Off farm enterprises				

Uses and desirable characteristics of product including uses of all parts of the plant or animal: a. consumption b. storage for later consumption, exchange, or sale c. other domestic use (e.g. fuel, building material) d. exchange e. sale f. reinvestment in agricultural production (e.g. manure) g. other

4. Inclusion Analysis

Inclusion analysis is a technique within the methodology of Farming Systems Research and Extension (FSR/E). It emphasizes the importance of farmers in FSR/E. Understanding a farming system and its practices requires involvement from all significant participants—those who perform the work, invest resources, and use the products. Inclusion analysis seeks to identify who participates at each stage or in each activity related to farming systems research and extension. Inclusion analysis is a technique within the methodology of Farming Systems Research and Extension (FSR/E). It emphasizes the importance of farmers in FSR/E. Understanding a farming system and its practices requires involvement from all significant participants—those who perform the work, invest resources, and use the products.

Inclusion analysis seeks to identify who participates at each stage or in each activity related to farming systems research and extension. To delve deeper into inclusion analysis, several key questions should be addressed:

A. What type of inclusion?

1. How are women and men included in the information gathering process?
2. In what capacities are they involved, such as being information sources, implementors, decision-makers, or beneficiaries?
3. Whose interests or preferences are considered and represented at each stage?

2. What are the criteria for including individuals?

What are the criteria for including individuals?

Is the selection of participants random or purposeful?

If non-random, what is the rationale behind the selection criteria?

It's noted that criteria for choosing farmers can sometimes unconsciously introduce biases, such as proximity to roads, endorsement by the village head, or membership in an organization limited to male heads of households. There might also be instances where the person performing a task is overlooked when collaboration is confined to the household head. The selection of farmer cooperators has been identified as a potential weak point in on-farm experimentation.

3. What steps are taken to encourage inclusion?

1. The mechanisms or methodology for collaborating with farmers play a crucial role in influencing participation and the quality of responses.
2. Considerations include the timing and frequency of visits or meetings, location, rules and means of access, whether interviews are conducted jointly, individually, or in groups, the focus of the questions asked, the level of flexibility or open-endedness in the process, the attitude of the researchers, and sometimes, the gender of the researchers.
3. Attention to these criteria and mechanisms is equally important when organizing the extension of promising technologies and supplying new inputs. Merely stating "open access" does not guarantee full participation or response; explicit efforts are needed to ensure meaningful engagement.

   In essence, this analysis aims to ensure that both men and women are effectively involved in the project, and the selection criteria and methodologies are fair, promoting inclusive and comfortable participation.

Table 4 is used to assess how are men and women included in FSR/E activities? What's the profile of their involvement? This table can be used to create a detailed overview of how both genders are included in research and extension activities. Assess this profile to identify gaps and improve the planning and monitoring of activities.

Table 4: Inclusion Analysis

Stages of FSR	Who is Included	Criteria for Inclusion	Mechanism of Inclusions
Diagnosis			
Planning and Design			
Experimentation and Evaluation			
Recommendations			
To researchers To Policy Makers To Extension			
Extension Information Input Credit Market Outlets			

Practical Tools of Gender Analysis to integrate gender considerations into livestock programming

USAID Gender and Livestock Brief has given useful tools to integrate gender considerations into livestock programming. The Livestock Producer Ownership Patterns Worksheet is a tool that can be used to identify the ownership patterns of livestock within a community. This information can be used to better understand the gender dynamics of livestock ownership and control, and to design interventions that promote gender equity and women's empowerment in livestock programming. Household Level Production Analysis Worksheet is a tool that can be used to guide a gendered analysis of the division of labour in the household by economic level. The purpose of this tool is to identify the different tasks and responsibilities associated with livestock production and management, and to understand how these tasks are divided between men and women within the household. The worksheet is designed to help practitioners identify shared responsibilities and coping mechanisms during times of peak labour needs and labour shortages due to illness or death. This information can be used to design interventions that promote more equitable and sustainable livestock production systems, and to ensure that women's contributions to livestock production are recognized and valued. The Gendered Livestock Stakeholder Matrix is a tool that can be used to map the key stakeholders within the target livestock system and the gender roles and relations across the prioritized empowerment domains. This tool can help identify the different actors involved in livestock production and marketing, and their respective roles and responsibilities. It can also help identify the gender-specific constraints and opportunities that exist within the livestock system and inform the design of interventions that promote gender equity and women's empowerment in livestock programming.

These tools are practical and easy to use and can be adapted to different contexts and livestock systems. They are designed to help practitioners integrate gender considerations into livestock programming, and to promote more equitable and sustainable livestock development outcomes.

Tool A: Livestock Producer Ownership Patterns Worksheet

This tool is designed to facilitate the examination of current patterns of access and control among livestock producers, with a breakdown based on economic status, gender, and type of livestock. The goal is to understand how different groups, particularly women and men from various economic backgrounds, engage with livestock program activities. The interpretation of findings will be shaped by the varying degrees of control held by women and men.

Wealth Group	Sex	Micro stock	Poultry	Small Ruminants	Cattle	Camel Traction
Very Poor	Female					
Very Poor	Male					
Poor	Female					
Poor	Male					
Middle	Female					
Middle	Male					
Richer	Female					
Richer	Male					

Note: The categories represent different wealth groups, and the cells can be filled with information regarding the ownership patterns of each group for various livestock species, including micro stock, poultry, small ruminants, cattle, and camel traction. This information will aid in assessing disparities in access and control among different segments of the community.

Tool B: Household Level Production Analysis Worksheet

This tool is designed to guide a gendered analysis of labour division within the household, considering economic levels. As you fill out this tool, also identify shared responsibilities and strategies employed during peak labour demands or shortages due to illness or death.

Livestock Species:	Middle Wealth HHs	Difference between middle and richer wealth HHs	Difference between middle and poorer wealth HHs
Production Activities	Women	Men	Girl/Boy/Other*
Feeding			
Watering			
Cleaning/Hygiene			
Grazing/tethering/herding			
Breeding			
Housing			
Fodder production/ collection			
Medication/Sick Animal Care			
Milking			
Egg collecting			
Slaughtering			
Processing/Value adding**			
Selling live animals			
Selling animal products			
Other			

*Note: Fill in the table with information regarding production activities undertaken by women, men, and others in households of different wealth levels. Additionally, describe coping mechanisms during peak labor needs and how the household is impacted if a key adult falls ill or passes away.

Tool C: Gendered Livestock Stakeholder Matrix

This tool aims to visualize key stakeholders in the targeted livestock system and understand gender roles and relations across prioritized empowerment domains.

Note: This matrix helps in identifying and analysing key stakeholders in the livestock system. It considers various aspects such as laws, policies, cultural norms, gender roles, access to resources, and decision-making patterns. The stakeholders listed can be expanded and refined based on specific contexts.

Stakeholder	Laws, Policies, Regulations, Institutional Practices	Cultural Norms and Beliefs	Gender Roles, Responsibilities, and Time Use	Access to and Control over Assets and Resources	Patterns of Power and Decision-making
Livestock producer					
Producer associations, farm-based organizations, and cooperatives					
Parent stock breeders, young stock providers					
Animal health input suppliers and veterinary service providers					
Animal feed and fodder producers and suppliers					
Extension and market information service providers					
Financial services (including BDS)					
Livestock traders					
Transporters					
Primary and secondary processors (e.g. abattoirs, dairies, hides, cheese makers, and restaurants)					
Exporters and importers					
Household level food preparers and consumers of animal source foods					

Source: USAID. (n.d.). Gender and Livestock Brief. Retrieved from https://www.usaid.gov/sites/default/files/documents/1860/Gender-and-Livestock-Brief.pdf

Guidelines for conducting gender analysis in livestock sector

Blum *et al.*, 2020 have suggested certain guidelines which have been modified to conduct gender analysis in livestock sector:

i. Gain an understanding of gender relations, the division of labour between men and women (who does what work), and who has access to, and control over, resources. For example, in the context of animal husbandry, understand the roles and responsibilities of men and women in livestock management. Identify who is primarily involved in tasks such as feeding, breeding, healthcare, and marketing.

ii. Include domestic (reproductive) and community work in the work profile

iii. Recognize the ways women and men work and contribute to the economy, their family and society

iv. Use participatory processes and include a wide range of female and male stakeholders at the governmental level and from civil society, including women's organizations and gender equality experts; engage both men and women in the community in participatory processes to gather insights on their perspectives and experiences in animal husbandry. Include representatives from women's organizations, gender equality experts, and relevant governmental and civil society stakeholders in the planning and decision-making processes.

v. Identify barriers to women's participation and productivity (social, economic, legal, political, cultural...); For example, explore social, economic, legal, political, and cultural barriers that may limit women's active participation in animal husbandry. This could include unequal access to resources, discriminatory practices, or restrictive cultural norms. Identify ways to address and mitigate these barriers.

vi. Gain an understanding of women's practical needs and strategic interests, and identify opportunities to support both; For example, understand the practical needs of women in animal husbandry, such as access to training, resources, and markets. Also, identify strategic interests, like opportunities for leadership or entrepreneurship, which can empower women in the long term.

vii. Consider the differential impact of the initiative on men and women and identify consequences to be addressed; analyse how any proposed initiatives or interventions in animal husbandry may differentially affect men and women. This could involve assessing changes in workload, access to resources, or income-generation opportunities.

viii. Establish baseline data, ensure sex-disaggregated data, set measurable targets and identify expected results and indicators; and Collect baseline data on the status of gender dynamics in animal husbandry. Ensure sex-disaggregated data on aspects such as the number of male and female farmers, their roles, and income levels. Set measurable targets to track progress in promoting gender equity in the sector.

ix. Outline the expected risks (including backlash) and develop strategies to minimize these. Anticipate potential risks, including backlash or resistance to gender-sensitive initiatives. Develop strategies to minimize these risks, such as community awareness programs, stakeholder engagement, and ongoing dialogue to address concerns.

Key issues to be considered when integrating gender issues with Stakeholders (Cloverson,2015)

The Six W Approach for Gender Mainstreaming in Veterinary and Animal Husbandry Extension involves a comprehensive analysis of the presence, roles, tasks, timing, locations, and reasons influencing the participation of men and women in extension activities.

1. Who is present or absent?

Evaluate the attendance of men and women, considering age groups and socio-cultural backgrounds during animal husbandry workshops and training programmes. A skewed representation may signal gaps in reaching a significant portion of the target population.

2. Who performs which tasks?

Identify gender-specific roles in animal husbandry, such as women's involvement in small-scale poultry or dairy activities and men's focus on larger livestock management. Understanding these roles informs tailored extension services to address the distinct needs and challenges faced by each gender.

3. What activities are undertaken?

Determine the specific tasks carried out by men and women in animal husbandry, such as women handling feeding and care, and men overseeing breeding or marketing. Recognition of these activities is essential for designing interventions aligned with existing gender dynamics.

4. When are activities conducted?

Consider the timing of animal husbandry activities concerning daily and seasonal responsibilities. When planning workshops, account for these factors to ensure they do not conflict with women's household duties, potentially impacting their ability to participate.

5. Where are activities conducted?

Recognize the locations where various animal husbandry tasks occur, such as women being more involved in on-farm activities and men engaging in off-farm tasks like marketing. Understanding these locations facilitates the design of accessible and relevant extension services for both genders.

6. Why are activities undertaken or not?

Investigate the underlying reasons for men's and women's participation or non-participation in animal husbandry extension activities. Factors like cultural norms, resource access, or time constraints should be understood to develop gender-sensitive advisory services addressing the specific constraints and opportunities faced by male and female farmers.

In summary, applying the six Ws to animal husbandry extension and advisory services helps ensure that programs are inclusive, culturally sensitive, and tailored to the unique roles and responsibilities of both men and women in the context of livestock management **and agriculture.**

Useful Terms in Gender Analysis

Tokenism:

It involves appointing a single woman to a committee or board primarily to represent the 'women's perspective'.

Gender statistics

They are defined a field of statistics which cuts across the traditional fields to identify, produce and disseminate statistics that reflect the realities of the lives of women and men and policy issues relating to gender equality.

Gender bias

Gender bias denotes an approach that, though often appearing neutral, subtly favours one gender (typically men) or harbours prejudice against another gender (typically women).

Gender blindness

Gender blindness involves attempting to address inequities by treating all individuals in a simplistic manner yet fails to acknowledge the gender implications and potential unintended consequences for people of different genders.

Gender role stereotyping is entails deeply ingrained and inflexible expectations and beliefs regarding the roles of women and men, such as the notion that 'women are responsible for household and childcare duties' while 'men are the primary earners'.

Intersectionality

Gender operates as a social variable that intersects with various other social variables, including age, ethnicity, class, religion, disability, sexual orientation, and more. Intersectionality highlights the dynamic interplay among these social variables and emphasizes that individuals exist at the intersection of multiple identities. For instance, a woman's identity is not singularly defined by gender; rather, it encompasses additional factors such as ethnicity, age, sexual orientation, and so forth.

Need for gender analysis

Gender' is a way of looking at how men and women's roles and responsibilities relate to each other. It's not about biological differences but about differences influenced by society and economy. Since not everyone in a household has equal access to resources or benefits from what's produced, ignoring gender issues can lead to project failures. Gender analysis, which looks at how men, women, and children are involved in farming, has become a common term. It helps projects work better and avoid problems. For example, if a new technology increases the total household income, it shouldn't make women's work much harder.

Gender analysis is an essential tool for incorporating a gender perspective into policies, programs, and projects. By delving into gender analysis, we can uncover the differences and inequalities between women and men, revealing their distinct roles, positions in society, resource access, opportunities, constraints, and power distribution in a specific context. This process forms the foundation for developing interventions that address these disparities and cater to the diverse needs of both women and men.

A comprehensive gender analysis equips policymakers with a deeper understanding of gender inequalities within a particular sector or situation. It goes beyond just describing the current gender-related conditions; it delves into the underlying causes and repercussions of these gender disparities on the targeted groups.

By examining the root causes of gender inequality and discrimination, it becomes possible to establish relevant, targeted objectives and measures to eliminate these disparities. In essence, gender analysis contributes to enhancing the gender responsiveness of policies and legislation, ensuring that the needs of all citizens, regardless of gender, are adequately addressed.

When applied to organizations and institutions, gender analysis also serves to unveil how the nature of service delivery may affect women and men differently or how institutions themselves exhibit gender biases. This is evident, for instance, in the workplace regarding recruitment practices, gendered divisions of labour, and women's access to decision-making roles.

In essence, gender analysis is a valuable tool for understanding the activities undertaken by women and men in society, along with the challenges and opportunities each group encounters. Its practical benefits include:

i. **Preventing Assumptions**: Gender analysis discourages making assumptions about the lives of women and men, girls and boys, instead emphasizing the need to comprehend their distinct needs, roles, social status, resource access, interests, capacities, power dynamics, and priorities.

ii. **Root Cause Understanding**: It seeks to fathom the underlying reasons behind these differences, providing a thorough understanding of an issue or situation while considering all segments of the population.

iii. **Highlighting Disadvantages:** Gender analysis uncovers how cultural, economic, and legal factors disadvantage women and girls, as well as men and boys, by impacting their opportunities throughout their lives. It explores the connections between inequalities at various societal levels.

iv. **Identifying Barriers**: It reveals how these differences can hinder women, girls, men, and boys from participating in or benefiting from programs or projects.

v. **Recommendations for Equity**: Gender analysis leads to specific recommendations for addressing the needs of women, men, girls, and boys equitably. This includes tackling gender discrimination, gender-based violence, and discriminatory gender norms.

vi. **Progress Evaluation**: Gender analysis is vital for monitoring and evaluating progress in closing the gender gaps in accessing and benefiting from interventions while also reducing gender discrimination.

Need for Gender Analysis in the Livestock Sector

Gender analysis in the livestock sector is crucial for various reasons. It allows us to uncover and address the unique roles, challenges, and opportunities faced by men and women involved in animal husbandry and related activities. It is essential for promoting equitable and sustainable development in the sector. Research has shown that women farmers often benefit the least from livestock extension services and technologies designed to improve production. In some cases, these technologies create additional burden for women. For instance, the introduction of mechanization, which primarily benefits men in tasks like land clearing and preparation, can lead to increased labor demands for planting, weeding, and harvesting, predominantly carried out by women.

Additionally, the underrepresentation of women in farming has resulted in missed opportunities to tap into their rich knowledge and experience in livestock management.

Gender analysis in the livestock sector serves the following purposes:

i. **Identifying Disparities**: It recognizes disparities in roles, access to resources, and decision-making power between men and women involved in livestock farming.

ii. **Addressing Challenges**: It helps develop strategies to overcome challenges faced by women, such as providing training at convenient times, ensuring access to credit, and promoting an equitable division of labor. iii. **Promoting Equality**: It encourages initiatives that engage men in sharing household and childcare responsibilities, giving women more time to participate in livestock farming activities and training.

iv. **Enhancing Productivity**: Involving women in decision-making and providing them with equal access to resources, can boost the productivity and profitability of livestock farms.

v. **Empowerment**: Enabling women to participate fully in the livestock value chain contributes to their economic empowerment and overall well-being.

vi. **Efficiency in design and implementation of project design** :Gender analysis helps project planners and implementers to identify the distinct roles and contributions of men and women in each setting. By understanding these roles, project interventions can be tailored to maximize efficiency. A gender analysis in animal husbandry would involve understanding the existing roles of men and women in livestock management. If women are primarily responsible for feeding and caring for the livestock, the efficiency of the project would depend on how well the new technology aligns with these existing roles. For instance, the technology should be user-friendly and not overly complex, ensuring that women can easily integrate it into their daily tasks without significant disruptions.

vii. **Reducing Negative Side-Effects:** Gender analysis is crucial for anticipating and minimizing unintended negative consequences that may disproportionately affect one gender. For example, if a technology increases the overall yield of a crop but also adds to the workload and drudgery for women, it could have adverse effects on their well-being and productivity. Similarly, if the introduction of the new feed supplement requires additional time, effort, or skills that disproportionately affect women's workload, it could lead to negative side-effects. For example, if the new supplement needs special preparation or handling, women may face an increased burden in managing the feeding process. Gender analysis helps identify such potential challenges and allows project planners to devise strategies to mitigate them.

viii. **Balancing Workload and Benefits**: Gender analysis ensures that the benefits of increased livestock productivity and income are shared equitably between men and women. If the extra income generated from selling livestock products is controlled by men without consideration for the increased workload borne by women due to the new technology, it could create gender disparities. To address this, the project could incorporate measures to empower women in decision-making regarding livestock income and ensure fair distribution of responsibilities.

In conclusion, gender analysis in the livestock sector, as illustrated in dairy farming, is essential for identifying and addressing gender disparities, advancing equitable development, and maximizing the sector's potential for all individuals involved. Thorough gender analysis, project designers can ensure that the introduction of new technologies in animal husbandry not only enhances efficiency and income but also considers the gender-specific implications, aiming for a more balanced and inclusive impact on both men and women involved in livestock management. Such analyses are vital steps toward achieving sustainable and inclusive agricultural practices.

Summary

- Gender analysis in livestock farming is crucial for understanding disparities between men and women in the sector.
- It reveals differences in roles, resource access, and decision-making power, providing insights into the specific challenges faced by each gender.
- This understanding allows for the development of tailored interventions to address gender disparities effectively.
- Promoting gender equality in livestock farming can lead to increased productivity and efficiency in the sector.
- It can also contribute to the economic empowerment of women who play a significant role in agriculture.
- Gender equality improves overall well-being, including living conditions, income, and food security.
- Furthermore, it promotes inclusivity and aims to eliminate gender discrimination in the sector.

- This analysis helps ensure that both men and women have equal opportunities to benefit from and contribute to livestock farming.
- It is essential for sustainable and equitable development within the livestock sector.
- Ultimately, gender analysis fosters economic empowerment, inclusivity, and better well-being for all involved in livestock farming.

1. Why is gender analysis essential in the livestock sector, particularly in animal husbandry? Provide an in-depth explanation of the significance of understanding and addressing gender disparities in livestock farming.

Gender analysis is crucial in the livestock sector, specifically in animal husbandry, because it sheds light on the differences and inequalities between men and women in this domain. It plays a vital role in understanding, acknowledging, and addressing these disparities.

In the context of animal husbandry, gender analysis delves into the specific roles, responsibilities, and access to resources and decision-making power of men and women. These insights are essential to recognizing and addressing the unique challenges and opportunities experienced by each gender.

For example, it reveals that men often dominate tasks like herding and decision-making, while women may primarily engage in activities like milking and domestic chores. This division of labour results in unequal access to income-generating opportunities and resources, disadvantaging women. By conducting gender analysis, these disparities can be identified and rectified.

Moreover, gender analysis extends to explore how broader societal norms, cultural practices, and economic systems intersect with gender roles. For instance, it can unveil how the gender wage gap within the livestock sector is connected to national policies, cultural expectations, and even family dynamics.

By promoting gender equality through effective gender analysis, the livestock sector can potentially reap numerous benefits. These include increased productivity, economic empowerment for women, and overall well-being within communities. Furthermore, it can serve as a tool to eliminate gender discrimination in various sectors and organizations, contributing to a more equitable and sustainable development approach.

2. What are the core aspects of gender analysis, and how do they contribute to a comprehensive understanding of gender disparities in various sectors, including animal husbandry?

Gender analysis comprises several critical aspects that provide a comprehensive understanding of gender disparities in sectors like animal husbandry. These aspects are:

i. Political and Legal Framework

a) Gender-Responsive National Policies and Laws: This aspect involves assessing whether national policies and regulations acknowledge gender equality and women's rights. It aims to determine the extent to which legal frameworks promote gender equity.

b) International Commitments: Evaluation of whether national policies align with international commitments on gender equality, such as the United Nations Sustainable Development Goals (SDGs).

In the context of animal husbandry, these aspects would assess whether national policies and laws related to livestock farming adequately consider gender equality and women's rights. It also checks if women have sufficient access to government support programs and resources, aligning with international commitments like the SDGs.

ii. Access to and Control over Resources

i. Resource Definition: Recognizing that resources encompass both material and intangible assets, including time, information, knowledge, and rights.

ii. Access to Resources: Investigating who has access to these resources and who benefits from them.

iii. Control over Resources: Analysing who could make decisions about resource utilization.

iv. Awareness of Legal Rights: Assessing whether women and men are aware of their legal rights regarding resource access and control.

In the context of animal husbandry, this aspect highlights the access of men and women to financial resources, livestock-related training programs, and the ability to make decisions about livestock management. It identifies whether the access and control over resources are unequal and how this imbalance affects women's participation in animal husbandry.

iii. Access to Services and Institutions

i. Availability and Equality: Evaluating the availability of services and institutions for women and men.

ii. Treatment by Service Providers: Examining whether there is bias or discrimination in service provision.

iii. Affordability: Analysing whether services are affordable for both genders.

Gender analysis in animal husbandry might reveal disparities in access to veterinary services, indicating that services are primarily available during working hours, which is inconvenient for women with domestic responsibilities. This aspect would also highlight any gender biases in service provision and assess affordability issues.

iv. Women's and Men's Roles

Productive Work: Investigating gender roles in income-generating activities and identifying whether there's horizontal or vertical segregation.

Household (Reproductive) Work: Analysing the division of household tasks, particularly unpaid care work.

Community Work: Assessing involvement in community activities and roles in various organizations.

In the context of animal husbandry, this aspect reveals the division of labour between men and women, with men primarily responsible for herding and decision-making, while women focus on milking and domestic tasks. This leads to unequal access to income-generating opportunities for women.

v. Participation in Decision-Making

Who Participates: Identifying individuals involved in decision-making processes.

Decision-Making Procedures: Analysing the methods used for decision-making.

Gender analysis might unveil that decision-making in the community's livestock management is predominantly male-dominated, and decision-making procedures don't consider the perspectives of women effectively.

These core aspects provide a holistic view of gender disparities, allowing for informed policy, program, and project development to address these disparities and promote gender equality.

3. What are the steps involved in conducting a gender analysis, and how can these steps lead to more gender-responsive and effective policies and programs in the context of animal husbandry?

The process of conducting a gender analysis involves several key steps, which, when followed, can lead to more gender-responsive and effective policies and programs, especially in animal husbandry:

a) Collection of Gender- and Sex-Disaggregated Data

The first step in conducting a gender analysis is the collection of gender- and sex-disaggregated data, both quantitative and qualitative. This data enables a project team to understand the need for gender mainstreaming activities.

Example in Animal Husbandry: The project team recognizes the need to collect data on how men and women are involved in livestock management. This includes quantitative data like the number of animals cared for by each gender and income generated. Qualitative data is collected through focus group discussions with women and men in the community. The team also consults with local stakeholders.

b) Identifying Gender Differences and Underlying Causes

A robust gender analysis identifies gender differences and the underlying causes of these inequalities, allowing for tailored interventions.

Example in Animal Husbandry: Through gender analysis, the project team may identify that women's limited access to veterinary services is a significant gender difference and a cause of gender inequality in animal husbandry. Men often make decisions about seeking veterinary care for livestock, leading to delayed or inadequate healthcare. Understanding this difference enables interventions to address the specific needs of women.

c) Informing Development of Policies, Programs, and Projects

The insights gained from gender analysis inform the development of policies, programs, and projects tailored to the needs of both women and men. This ensures that these initiatives are more credible and valid among those affected by them.

Example in Animal Husbandry: Building on gender analysis findings, the project team advises local government authorities and agricultural development organizations to address gender disparities. They present evidence of gender-based differences in livestock care and unequal access to veterinary services. In response, the local government implements a policy ensuring equal access to veterinary services. Additionally, they design a training program specifically tailored to women's needs in the community, aiming to improve animal healthcare and household income.

By following these steps, a gender analysis can significantly enhance the responsiveness and effectiveness of policies and programs, addressing the specific challenges faced by both women and men in animal husbandry and promoting greater gender equality within the sector.

4: What are the key differences between practical and strategic gender needs in the livestock sector, and how can understanding these distinctions improve gender-responsive interventions?

Practical gender needs in the livestock sector refer to the everyday requirements and challenges faced by both women and men. For example, a practical gender need could involve women needing better access to veterinary training or improved access to water sources for livestock. These needs address immediate concerns but do not challenge existing gender roles.

On the other hand, strategic gender needs have the potential to transform existing gender inequalities. In animal husbandry, this could involve challenging traditional gender roles where men make major decisions related to livestock breeding and healthcare. Women may need access to knowledge and resources that enable them to participate in decision-making about livestock management. Understanding the difference between these two types of needs allows for more targeted interventions. Strategic gender needs are crucial for achieving long-term gender equality in the livestock sector.

5: How does the Moser conceptual framework for gender analysis and planning differentiate between practical gender needs and strategic gender needs, and why is this differentiation significant in promoting gender equality in livestock farming?

The Moser conceptual framework distinguishes between practical gender needs and strategic gender needs. Practical gender needs are immediate, everyday needs related to the daily lives of women and men in the community. For example, women may need access to training in modern milking techniques to improve their daily tasks. These needs address daily challenges but may not challenge existing gender roles. In contrast, strategic gender needs have the potential to transform existing gender subordination. In the livestock sector, this could involve women gaining access to knowledge and resources that enable them to participate in decision-making about livestock management.

This differentiation is significant because it helps in tailoring interventions. Understanding the specific needs of women in livestock farming, whether practical or strategic, allows for the development of targeted programs and policies that address these needs. By addressing both practical and strategic gender needs, interventions can work towards more equitable gender relations, ultimately improving the overall well-being and productivity of the livestock sector.

6. How does the social relations approach contribute to understanding gender disparities in small-scale poultry farming, and what role do institutions play in perpetuating or challenging these disparities?

The social relations approach contributes to understanding gender disparities in small-scale poultry farming by examining the structural connections that perpetuate systemic gender inequalities. In small-scale poultry farming, this approach reveals how gender roles and responsibilities create disparities in workload and decision-making. For example, women may be responsible for daily tasks like feeding and egg collection, while men handle larger tasks such as building poultry housing structures. By analysing these social relations, we gain insights into the dynamics at play.

Institutions in the livestock sector, including government agencies, NGOs, and local cooperatives, have a significant influence on gender disparities. Gender biases within these institutions can reinforce existing inequalities. For example, if training programs primarily target men due to traditional gender biases, it perpetuates the unequal division of labour and decision-making. Understanding how institutions interact with farmers helps identify areas where change is needed to challenge and rectify gender disparities.

7. What are the four types of gender policies within the 4R method, and how can they influence gender equality in livestock cooperatives?

The 4R method for gender analysis includes four types of gender policies:

i. **Gender-Blind Policies:** These policies consider men and women unequal and often reinforce traditional gender norms, roles, and stereotypes. For instance, a gender-blind policy might provide subsidies for poultry feed without considering the different roles of men and women in poultry farming, which can reinforce existing gender inequalities.

ii. **Gender-Neutral Policies:** Gender-neutral policies operate within existing gender divisions of resources and responsibilities without challenging them. These policies assume that men and women are not equal but the same. A gender-neutral policy might provide equal access to resources without recognizing the different roles of men and women. This policy doesn't challenge existing gender norms and may leave workload and decision-making dynamics unchanged.

iii. **Gender-Sensitive Policies:** Gender-sensitive policies consider men and women as equals and address gender norms, roles, and resource access to achieve policy objectives. For example, a gender-sensitive policy might provide training and resources to women in poultry farming to improve their skills and knowledge, thereby promoting more equitable gender relations within the sector.

iv. **Gender-Positive Policies:** Gender-positive policies also assume men and women as equals and focus on changing gender norms, roles, and access to resources as core policy outcomes. A gender-positive policy could actively promote women's participation in poultry farming, provide leadership opportunities, and challenge traditional gender norms, leading to greater empowerment and economic independence for women in the sector.

These policies can influence gender equality in livestock cooperatives by determining how resources, training, and opportunities are distributed among men and women. Gender-sensitive and gender-positive policies aim to promote gender equality by recognizing and addressing the specific needs and challenges faced by women, thus ensuring more equitable and inclusive cooperative environments.

SECTION B: SHORT ANALYTICAL QUESTIONS

1. **What is the purpose of a gender analysis?**

   The purpose of a gender analysis is to systematically examine and understand the differences and disparities in roles, access to resources, decision-making, and power between women, men, girls, and boys within a specific context. It aims to provide insights into the dynamics of gender-related norms, such as identity and sexuality, and inform policies and programs to promote gender equality and inclusivity.

2. **What does a comprehensive gender analysis encompass?**

   A comprehensive gender analysis includes the following critical aspects:

   Political and Legal Framework: Assessing the alignment of national policies and laws with international commitments on gender equality.

   Access to and Control over Resources: Examining the distribution and control of material and intangible resources among different gender groups.

   Access to Services and Institutions: Evaluating the availability, affordability, and treatment by service providers for both women and men.

   Women's and Men's Roles: Analyzing gender roles in productive work, household work, and community work.

   Participation in Decision-Making: Identifying who participates in decision-making processes and the procedures used.

3. **Why is it important to consider gender in the analysis of access to resources and services?**

   Considering gender in the analysis of access to resources and services is crucial because it helps identify and address disparities in resource allocation, access, and treatment between women and men. Gender analysis ensures that policies and programs are designed to be equitable and responsive to the specific needs and challenges faced by different gender groups.

4. **What can a gender analysis reveal in the context of animal husbandry?**

   In the context of animal husbandry, a gender analysis can reveal differences in roles, responsibilities, and access to resources between women and men. It may highlight inequalities in livestock care, decision-making, and access to services. For example, it can show that women are primarily responsible for small livestock care, while men are more involved in herding larger animals, leading to disparities in livestock management.

5. **How does a comprehensive gender analysis contribute to the development of policies and programs?**

   A comprehensive gender analysis provides valuable insights into the dynamics of gender roles, rights, and disparities. It informs the development of policies, programs, and projects that are more equitable, responsive, and inclusive of both women and men. For example, it can lead to the design of programs that provide women with access to veterinary services outside regular working hours, training opportunities in animal healthcare, and more equitable participation in decision-making processes, promoting gender equality within the livestock sector.

6. **What is gender analysis, and why is it important in various contexts?**

   Gender analysis is the systematic study of differences in the roles, responsibilities, access to resources, and power between women, men, girls, and boys within a given context. It is essential because it helps identify and understand these gender-based differences, leading to more inclusive and equitable policies, programs, and projects.

7. **What are the key steps involved in conducting a gender analysis**?

   There are three key steps in conducting a gender analysis:

   Collection of Gender- and Sex-Disaggregated Data

   Identifying gender differences and underlying causes of gender inequalities

   Informing the development of policies, programs, and projects that respond to the different needs of women and men.

8. **How does a political and legal framework relate to gender analysis, and why is it important?**

   The political and legal framework encompasses policies and laws that can either promote or hinder gender equality. It is important because it assesses whether national laws and policies align with international commitments on gender equality and women's rights, ensuring that gender disparities are addressed.

9. **What does a gender analysis reveal about access to and control over resources?**

   A gender analysis highlights disparities in the access to resources and who benefits from them. It examines factors that influence control over resources, including legal rights and awareness. This analysis is crucial in understanding how gender inequalities are perpetuated.

10. **How does gender analysis relate to roles in productive, household, and community work?**

    Gender analysis examines the division of roles between women and men in productive work, household (reproductive) work, and community work. It helps identify inequalities, such as vertical or horizontal segregation, which impact access to income-generating opportunities for women.

11. **What is the significance of gender analysis in participation in decision-making processes**?

    Gender analysis identifies who participates in decision-making processes and whether these processes consider the needs and perspectives of both women and men. It helps reveal disparities in influence and control, which can inform policies and programs for more equitable participation.

12. **In the context of animal husbandry, what tasks are primarily performed by women and what tasks by men?**

    In many cases, women are responsible for tasks like daily animal care, feeding, and milking, while men often handle tasks requiring physical strength and construction, such as building animal shelters or handling larger livestock.

13. **Who controls major decisions and financial resources in animal husbandry?**

    Men tend to make decisions related to significant financial transactions and investments, while women often oversee day-to-day tasks and expenses.

14. **What cultural norms and expectations affect gender disparities in animal husbandry roles and access to resources?**

    Traditional gender norms and expectations can reinforce the division of labor and decision-making in animal husbandry. For example, cultural norms may dictate that men have more decision-making authority in financial matters.

15. **How are gender roles identified in animal husbandry, particularly concerning daily tasks and responsibilities?**

    Gender roles in animal husbandry can be identified by examining who is responsible for specific activities, such as milking, feeding, and overall animal care.

16. **What are the practical and strategic gender needs in animal husbandry, and how can they be addressed?**

    Practical gender needs may involve access to resources like training and healthcare, while strategic gender needs might pertain to decision-making and participation in higher-profit activities.

17. **What factors influence the control of resources and decision-making in households engaged in animal husbandry?**

    Cultural norms and traditional gender roles often influence decision-making power and resource allocation, with men having more control over major decisions.

18. **How can the burden of the "triple role" faced by women in animal husbandry be alleviated?**

    Strategies such as introducing labor-saving technologies and involving men in household chores can help reduce the burden on women.

19. **What is the aim of specific policy interventions in animal husbandry, and how do they address practical and strategic gender needs?**

    Policies may focus on immediate needs like access to training or aim to promote women's leadership in collective farming initiatives.

20. **How can women and gender-aware organizations be engaged in the planning process to ensure interventions are tailored to women's specific needs?**

    Women's active participation is crucial in decision-making to ensure that interventions consider their unique challenges and requirements.

21. **How can the Social Relations approach contribute to the understanding of gender dynamics in small-scale poultry farming?**

    The approach can help uncover how gender roles and norms impact the distribution of tasks and decision-making in poultry farming, as well as how institutions and policies influence these dynamics.

22. **How does the gender analysis matrix framework assess labor, time, resources, and sociocultural factors in animal husbandry?**

    The framework examines how these factors are distributed between genders and identifies disparities in labor, resource access, and decision-making.

23. **What are the four steps of the 4R method, and how can they be applied in the context of a livestock cooperative?**

    The four steps involve representation, resource allocation, analyzing conditions, and realizing gender equality. In a livestock cooperative, this could mean examining the gender distribution in leadership positions, resource access, and the underlying reasons for disparities, and then formulating objectives and measures for achieving gender equality.

24. **Why is gender analysis important in the context of the livestock sector?**

    Gender analysis is crucial for identifying disparities, addressing challenges faced by women, promoting equality, enhancing productivity, and empowering women in animal husbandry. It ensures that the diverse needs and roles of both women and men are adequately considered in policies and interventions.

25. How does the 'Activity' profile in the Harvard Analytical Framework help us understand gender dynamics in animal husbandry?

The 'Activity' profile helps us identify who performs various tasks in animal husbandry, shedding light on gender-specific roles. For instance, it can reveal whether men are primarily responsible for physically demanding tasks while women handle tasks requiring gentler care, thereby highlighting gender-based divisions in responsibilities.

26. How can the 'Access' and 'Control' profiles of the Harvard Analytical Framework inform interventions in animal husbandry?

The 'Access' and 'Control' profiles help in identifying who has access to resources and decision-making power in animal husbandry. This information can guide interventions by addressing inequalities in resource allocation and promoting more inclusive and equitable access to benefits.

27. What role do 'Influencing Factors' play in the Harvard Analytical Framework?

'Influencing Factors' are essential for understanding the reasons behind gender disparities. They can include cultural norms and economic factors, shedding light on why specific gender roles and resource allocations exist.

28. How does the Moser Conceptual Framework distinguish between 'Practical Gender Needs' and 'Strategic Gender Needs' in animal husbandry?

The framework helps us differentiate between the immediate needs of women and men, such as access to training, and the long-term needs related to challenging traditional gender roles, such as women's participation in decision-making.

29. What is the significance of 'Planning for Balancing the Triple Role' in the Moser Framework in the context of animal husbandry?

This aspect of the framework addresses the challenge of women's triple role – reproductive work, productive work, and community management. It involves planning strategies to alleviate this burden, ensuring that women have time for income-generating activities and community involvement.

30. How does the Social Relations Framework view development in the context of animal husbandry?

The framework views development as an improvement in the well-being of both men and women involved in animal husbandry. It encompasses aspects like equitable access to resources, training, and opportunities related to livestock care and management.

31. How can an analysis of 'Social Relations and Structural Differences' impact gender equality in animal husbandry?

Analyzing social relations and structural differences can reveal how gendered roles and responsibilities create disparities in decision-making and resource access. Addressing these disparities is crucial for promoting gender equality.

32. How does the 'Representation' step in the 4R Method apply to an animal husbandry organization?

In animal husbandry organizations, 'Representation' would involve assessing the gender distribution at all levels of decision-making, such as leadership roles and staff positions. This provides insight into the gender balance within the organization.

33. Why is the 'Realia' step important in the 4R Method for gender analysis in animal husbandry?

'Realia' analysis helps identify the underlying reasons for gender disparities in representation and resource allocation. Understanding these reasons is crucial for designing effective interventions to achieve gender equality.

34. How can the 'Realisation' step in the 4R Method lead to positive changes in gender dynamics within an animal husbandry organization?

The 'Realisation' step involves formulating objectives and measures to achieve gender equality. It can lead to changes like inclusive decision-making processes, targeted training programs, awareness campaigns, and promoting economic empowerment, ultimately fostering more equitable gender relations within the organization.

35. Why is gender analysis considered crucial throughout the life of a development project?

Gender analysis is vital because it helps in understanding and addressing the roles, responsibilities, and dynamics between men and women in different project stages, ensuring that the project effectively meets the diverse needs and circumstances of both genders.

36. Provide an example of a project in India where gender considerations were incorporated later in the project. What was the impact of this inclusion?

The BIOCON project in India initially focused on treating straw for mixed farming involving both men and women. Later in the project, gender became a relevant aspect, leading to increased awareness and the collection of gender-segregated data. This inclusion highlighted the significant role of women in animal care and altered perceptions of technology in the context of users' living conditions.

37. What is the purpose of the Conceptual Framework for Gender Analysis in Farming Systems Research and Extension (FSR/E)?

The framework aims to correct gender bias in agricultural research and extension by recognizing the valuable contributions of all household members, including women, seniors, and children. It provides a guide for collecting and analysing information about gender roles and household dynamics within farming systems to design improved agricultural and livestock technologies.

38. In FSR/E, why is combining technical and social science data essential?

Combining technical and social science data is crucial for identifying problems, exploring technical possibilities for new technologies, and understanding farmers' decision-making processes. It helps align research areas with farmers' needs and ensures that the technologies developed are suitable for specific farming systems.

39. What factors shape decision-making in farming within societies?

Decision-making in farming is significantly influenced by factors such as gender differences, age, household position, class, ethnicity, and life cycle stages. Understanding these dynamics is essential for developing technologies that effectively address the diverse needs and circumstances of farmers.

40. What is the purpose of the "Activities Analysis" in farming systems research?

Activities Analysis aims to identify who performs specific tasks related to crop, livestock, household production, and other activities in farming systems. It helps in understanding gender roles and responsibilities, labor distribution, and potential impacts of proposed changes on different genders.

41. How does "Resources Analysis" contribute to farming systems research?

Resources Analysis helps in understanding who has access to and control of critical resources (e.g., land, water, labor) within a farming system. This analysis provides insights into decision-making processes, identifies production constraints, and informs the design of interventions by considering resource availability and control.

42. What is the goal of "Inclusion Analysis" in farming systems research and extension?

Inclusion Analysis aims to ensure that both men and women are effectively involved in the project. It assesses who is included at each stage, the criteria for inclusion, and the steps taken to encourage participation. The goal is to promote fair, inclusive, and comfortable participation for all individuals, regardless of gender.

43. How is the "Resources Analysis" conducted in FSR/E?

Resources Analysis involves disaggregating by gender and age who has access to and control of critical resources. It helps understand farm management decisions influenced by resource availability and control, providing a map for production constraints and proposed solutions.

44. What does "Benefits and Incentive Analysis" focus on in farming systems research?

Benefits and Incentive Analysis focuses on who has access to or control of the output of production and explores the preferences of users. It delves into how the output is used and examines incentives related to characteristics of the produce, improvements in yields or income, stability in yields, and reduced labor requirements.

45. Why is "Inclusion Analysis" important in farming systems research and extension?

Inclusion Analysis ensures that both men and women are effectively involved in the project, examining the criteria and mechanisms for their inclusion. It aims to identify gaps, promote fair selection, and encourage comfortable and inclusive participation of all stakeholders.

46. What are the key stages covered in the "Inclusion Analysis"?

The key stages in Inclusion Analysis include Diagnosis, Planning and Design, Experimentation and Evaluation, Recommendations to Researchers, Policy Makers, and Extension. It explores who is included, criteria for inclusion, and mechanisms of inclusion at each stage of FSR/E activities.

47. What is the key emphasis of the Activities Analysis in farming systems research, and how is it beneficial in project planning?

The key emphasis of Activities Analysis is to identify who performs specific tasks in farming, considering different seasons. It helps create an activities map to screen problems, select research priorities, designate collaborating farmers, and design on-farm trials.

48. **Explain the difference between access and control in the context of the Resources Analysis stage in FSR/E.**

    Access refers to the freedom or permission to use a resource, while control is the power to decide how a resource is used. Provide examples illustrating these concepts, especially in the context of farm management decisions.

49. **What is the significance of gaining an understanding of gender relations in livestock programming?**

    It is crucial for designing interventions that promote gender equity and women's empowerment in livestock development outcomes.

50. **Why is it important to include domestic and community work in the work profile, according to the guidelines?**

    It recognizes the full spectrum of work contributions by women and men to the economy, family, and society.

51. **What are some examples of barriers to women's participation in animal husbandry, as suggested by the guidelines?**

    Examples include social, economic, legal, political, and cultural barriers, such as unequal access to resources or discriminatory practices.

52. **What is the difference between practical needs and strategic interests of women in animal husbandry?**

    Practical needs involve immediate requirements like training and resources, while strategic interests focus on long-term empowerment opportunities.

SECTION C: OBJECTIVE QUESTIONS

Question 1. Multiple Choice Questions:

1. **What is the primary objective of gender analysis in animal husbandry?**
    a. To increase livestock productivity
    b. To uncover and address gender-based disparities
    c. To promote mechanization in animal farming
    d. To improve the quality of animal products

2. **What is the significance of collecting gender- and sex-disaggregated data in animal husbandry?**
    a. It helps identify the best breed of livestock.
    b. It ensures uniform resource allocation.
    c. It uncovers roles and responsibilities among women and men.
    d. It promotes equal access to veterinary services.

3. In gender analysis, what does "resource" encompass besides material and financial assets?

a. Marketing strategies

b. Intangible resources like time and knowledge

c. Government policies

d. Political affiliations

4. Why is the awareness of legal rights crucial for gender analysis?

a. To ensure women's domination in decision-making

b. To help enforce laws and regulations

c. To discourage women from pursuing livestock farming

d. To impose gender norms

5. What is the primary focus of a gender analysis in animal husbandry when examining access to services and institutions?

a. Availability of international organizations

b. Treatment by service providers

c. Socio-economic status

d. Affordability of luxury items

6. In the context of gender analysis in animal husbandry, what does "productive work" refer to?

a. Household chores

b. Selling livestock

c. Unpaid care work

d. Decision-making in the community

7. Why is unequal access to income-generating opportunities a common issue in animal husbandry?

a. Lack of resources

b. Gender-based disparities in roles

c. Excessive government intervention

d. Unequal access to technology

8. What can a gender analysis reveal about decision-making in the livestock sector?

a. It confirms women's dominance in decision-making.

b. It identifies challenges and disparities.

c. It suggests men's superiority in all areas.

d. It proposes reducing women's participation in decision-making.

9. How does a robust gender analysis contribute to the livestock sector?

a. By increasing gender-based disparities
b. By promoting unequal access to resources
c. By addressing gender disparities and promoting equality
d. By encouraging gender discrimination

10. What is the first step in conducting a gender analysis in animal husbandry?

a. Identifying gender differences
b. Informing development policies
c. Collection of gender- and sex-disaggregated data
d. Creating marketing strategies

11. What are some aspects of gender-disaggregated data in animal husbandry?

a. Best time to feed livestock
b. Land prices in rural areas
c. Roles and responsibilities in livestock care
d. International trade policies

12. What can an understanding of the control over resources reveal in gender analysis?

a. How to monopolize resources
b. How resources are allocated
c. Resource distribution among different genders
d. How to minimize resource utilization

13. In gender analysis, why is the affordability of services and institutions significant?

a. It ensures a monopoly of services.
b. It encourages higher prices for services.
c. It contributes to gender disparities.
d. It fosters equal access to services.

14. How does gender analysis address unequal access to veterinary services in animal husbandry?

a. By promoting male dominance in animal healthcare
b. By maintaining the status quo
c. By establishing mobile veterinary clinics
d. By discouraging women from seeking veterinary care

15. What is the ultimate goal of a gender analysis in the livestock sector?

a. To perpetuate gender-based disparities

b. To discourage women from participating in animal husbandry

c. To address gender disparities and promote equality

d. To reinforce gender discrimination

16. In gender analysis, what does "household (reproductive) work" typically encompass?

a. Income-generating work

b. Selling livestock

c. Unpaid care work

d. Decision-making in the community

17. Which framework examines gender dynamics within various activities, resource allocation, and control over benefits, emphasizing an 'activity' profile, 'access' profile, and 'control' profile?

a. Harvard Analytical Framework

b. Moser Conceptual Framework

c. Social Relations Approach

d. 4R Method

18. In the Moser Conceptual Framework, what distinguishes practical gender needs from strategic gender needs?

a. Practical gender needs address immediate challenges, while strategic gender needs focus on long-term transformation.

b. Practical gender needs involve access to resources, while strategic gender needs involve decision-making power.

c. Practical gender needs concern women, and strategic gender needs concern men.

d. Practical gender needs consider productive work, and strategic gender needs consider reproductive work.

19. The Social Relations Approach to gender analysis is based on several core concepts. Which one involves enhancing human well-being through activities like small-scale poultry farming?

a. Social Relations

b. Development

c. Institutional Analysis

d. Gender Policies

20. Gender analysis can distinguish between different types of gender policies. Which type of policy presumes men and women are equals and actively challenges traditional gender norms?

a. Gender-Blind Policies

b. Gender-Neutral Policies

c. Gender-Sensitive Policies

d. Gender-Positive Policies

21. **What does the 4R method stand for in gender analysis?**
    a. Representation, Resources, Realia, Realization
    b. Roles, Responsibilities, Resources, Relationships
    c. Resources, Roles, Relations, Representation
    d. Realia, Resources, Representation, Relations

22. **Which step of the 4R method involves surveying the gender representation within an organization?**
    a. Representation
    b. Resources
    c. Realia
    d. Realization

23. **What aspect does the Resources dimension of the 4R method examine?**
    a. Access and control of resources between women and men
    b. Distribution of labour between women and men
    c. Underlying reasons for gender distribution
    d. Development of new gender-sensitive policies

24. **In gender analysis, what is the "Realia" stage primarily focused on?**
    a. Identifying gender disparities
    b. Developing gender-sensitive policies
    c. Assessing the underlying reasons for gender distribution
    d. Formulating new objectives and measures for gender equality

25. **Gender analysis helps in:**
    a. Promoting traditional gender roles and stereotypes.
    b. Identifying challenges faced by women and developing strategies to overcome them.
    c. Reinforcing gender disparities and discrimination.
    d. Advancing gender inequalities in the workplace.

26. **Why is gender analysis essential in the livestock sector?**
    a. To reinforce traditional gender roles and responsibilities.
    b. To ensure women are primarily involved in decision-making.
    c. To uncover and address gender disparities and promote equitable development.
    d. To exclude women from the livestock value chain.

27. What are the three interconnected components of the Harvard Analytical Framework for understanding gender dynamics?

a. Gender-blind, gender-neutral, and gender-positive

b. Activity, access, and control

c. Representation, resources, and realia

d. Gender roles, gender norms, and gender sensitivity

28. In the Harvard Analytical Framework, the "activity" profile aims to answer which question?

a. "Who has control over resources?"

b. "Who does what?"

c. "What are the underlying causes of gender disparities?"

d. "What are the practical gender needs?"

29. The Moser conceptual framework for gender analysis distinguishes between which two types of gender needs?

a. Practical and theoretical gender needs

b. Basic and advanced gender needs

c. Practical and strategic gender needs

d. Every day and exceptional gender needs

30. What does the "Realisation" step in the 4R method involve?

a. Surveying gender representation

b. Analysing conditions and underlying reasons for gender disparities

c. Formulating new objectives and measures to achieve gender equality

d. Collecting data on resource allocation

31. Which type of gender policy actively challenges and transforms gender norms and aims to create more equitable opportunities and outcomes for all genders?

a. Gender-blind policies

b. Gender-neutral policies

c. Gender-sensitive policies

d. Gender-positive policies

32. Gender analysis helps in preventing assumptions about the lives of women and men by:

a. Encouraging gender stereotypes

b. Focusing on gender-blind policies

c. Recognizing distinct needs, roles, and opportunities

d. Perpetuating gender-based discrimination

33. In the livestock sector, what is one potential consequence of the introduction of mechanization primarily benefiting men?

a. Decreased labour demands for women

b. Reduced opportunities for women's participation

c. Equitable distribution of tasks

d. Increased women's representation in decision-making

34. Gender analysis in the livestock sector can contribute to women's economic empowerment by:

a. Reinforcing traditional gender roles

b. Providing training at inconvenient times

c. Ensuring equitable access to resources

d. Promoting gender discrimination

35. What is the primary purpose of the gender analysis matrix framework?

a. To maintain existing gender disparities

b. To challenge and transform gender norms

c. To address the needs of men and boys

d. To exclude women from decision-making

36. In farming systems research, what is the primary unit of analysis often overlooked in traditional projects?

a. Male heads of households

b. Adult women

c. Senior men and women

d. Children

37. Which stage of FSR/E involves the collection of agroclimatic, biological, and socioeconomic data for identifying research priorities?

a. Diagnosis

b. Planning and Design

c. Experimentation and Evaluation

d. Recommendations

38. What is emphasized in the conceptual framework for gender analysis in farming systems research and extension?

a. Crop yields

b. Interconnected farming systems

c. Livestock management

d. Socioeconomic data

39. What does the activities analysis under Gender analysis aim to determine?

a. Crop yields

b. Who does what tasks in farming

c. Resource availability

d. Incentives for farmers

40. Under Gender analysis What is the purpose of the Resources Analysis table?

a. Identifying problems in farming systems

b. Analysing gender roles

c. Describing farming calendars

d. Understanding resource access and control

41. Under Gender analysis What is the focus of the benefits and incentive analysis?

a. Gender roles

b. Crop yields

c. Output of production

d. Resource availability

42. Under Gender analysis Incentive analysis explores the motivations of farmers to:

a. Increase labor demands

b. Persist with existing practices or adopt new ones

c. Reduce environmental impacts

d. Ignore market appeal

43. Under Gender analysis What is the primary implication of the Inclusion Analysis stage?

a. Understanding gender roles

b. Encouraging farmer participation

c. Ensuring fair and inclusive participation

d. Identifying resource shortages

44. Under Gender analysis What does the Inclusion Analysis aim to ensure in FSR/E activities?

a. Random selection

b. Bias in participant selection

c. Fair and inclusive participation

d. Comfortable participation for men only

45. During which stage of FSR/E should socioeconomic data related to Gender be incorporated?

a. Diagnosis

b. Experimentation and Evaluation

c. Planning and Design

d. Recommendations

46. What is the primary emphasis of Farming Systems Research and Extension (FSR/E)?

a. Isolated crops or livestock

b. Inter connected farming systems

c. Seasonal calendars

d. Individual farmers

47. What does the conceptual framework for gender analysis aim to correct in agricultural research and extension?

a. Bias in favor of men

b. Bias in favor of women

c. Focus on isolated crops

d. Neglect of valuable contributions from other household members

48. Why is it important to consider the users' living conditions in technology implementation?

a. To increase workload

b. To complicate project dynamics

c. To align technology with users' needs

d. To ignore gender roles

49. Under Gender analysis what is the primary emphasis of the Resources Analysis table?

a. Crop yields

b. Identifying problems in farming systems

c. Resource access and control

d. Socioeconomic data

50. Under Gender analysis what does the Activities Analysis aim to create with the use of two worksheets?

a. Crop yields map

b. Resource availability map

c. Activities map or profile

d. Gender roles map

51. What is the purpose of the Livestock Producer Ownership Patterns Worksheet?

a. Assessing livestock market trends

b. Identifying key stakeholders

c. Understanding gender dynamics in livestock ownership

d. Analyzing household income levels

52. Which tool is designed to guide a gendered analysis of the division of labor within the household?

a. Livestock Producer Ownership Patterns Worksheet

b. Household Level Production Analysis Worksheet

c. Gendered Livestock Stakeholder Matrix

d. Gender and Livestock Brief

53. What is the focus of the Household Level Production Analysis Worksheet?

a. Livestock market trends

b. Division of labor within the household by economic level

c. Ownership patterns of livestock

d. Livestock stakeholder mapping

54. How can the Gendered Livestock Stakeholder Matrix be beneficial?

a. Identifying ownership patterns

b. Analyzing household income levels

c. Mapping key stakeholders and gender roles

d. Assessing livestock market trends

55. In the Livestock Producer Ownership Patterns Worksheet, what do the wealth groups represent?

a. Age groups

b. Income levels

c. Educational levels

d. Livestock species

56. What information does the Household Level Production Analysis Worksheet aim to collect?

a. Livestock market trends

b. Ownership patterns

c. Division of labor within the household

d. Key stakeholder mapping

57. According to the guidelines for conducting gender analysis, what should be included in the work profile?

a. Only economic work
b. Only community work
c. Domestic and community work
d. Only reproductive work

58. How can participatory processes be utilized in gender analysis?

a. By excluding stakeholders
b. By engaging a wide range of stakeholders
c. By focusing only on women's perspectives
d. By avoiding community participation

59. What is the purpose of identifying barriers to women's participation in animal husbandry?

a. To reinforce discriminatory practices
b. To assess livestock market trends
c. To address and mitigate obstacles
d. To prioritize men's roles

60. Which tool helps visualize key stakeholders and understand gender roles in the livestock system?

a. Livestock Producer Ownership Patterns Worksheet
b. Household Level Production Analysis Worksheet
c. Gendered Livestock Stakeholder Matrix
d. Gender and Livestock Brief

61. What is the significance of recognizing ways women and men contribute to the economy in gender analysis?

a. To prioritize men's contributions
b. To exclude women from economic activities
c. To understand diverse contributions
d. To promote inequality

62. According to the guidelines, why is it important to gain an understanding of women's practical needs and strategic interests?

a. To ignore women's contributions
b. To reinforce gender stereotypes
c. To promote long-term empowerment
d. To prioritize men's interests

63. According to the guidelines, what should be included in the baseline data for gender analysis?

a. Only male farmers' data

b. Current gender dynamics in animal husbandry

c. Only aggregate data

d. Data on livestock market trends

64. Why is it essential to establish baseline data in gender analysis?

a. To ignore changes in gender dynamics

b. To track progress in promoting gender equity

c. To prioritize male farmers

d. To avoid measurable targets and indicators

Question 2. True or False

1. Gender analysis explores differences between women, men, girls, and boys in resource distribution, opportunities, constraints, and power.
2. Gender analysis involves examining how responsibilities are divided among different genders in society.
3. Gender analysis focuses solely on the distinction between women and men.
4. Gender analysis encompasses quantitative data but not qualitative insights.
5. Gender analysis considers the specific individuals within gender categories.
6. Gender roles learned within families are rarely reinforced in the public domain.
7. Gender analysis doesn't explore the impact of gender norms on broader societal and cultural factors.
8. Gender analysis is concerned only with the roles of women and men in a society.
9. Gender analysis helps in understanding how different genders are involved in livestock management.
10. Gender analysis typically involves examining the status of women but not men in various aspects of life.
11. Gender analysis in animal husbandry does not consider the distribution of resources among different gender groups.

12. Gender analysis can identify the specific needs of women in accessing veterinary services.
13. Gender analysis does not take into account the quality of care received by women and men.
14. Gender analysis is not concerned with the affordability of services for women and men.
15. Gender analysis should assess whether national policies and laws acknowledge gender equality.
16. Gender analysis does not consider the impact of international commitments on gender equality.
17. Gender analysis evaluates the distribution of household tasks, including unpaid care work.
18. Gender analysis may reveal unequal access to income-generating opportunities for women in animal husbandry.
19. Gender analysis involves analyzing participation in community activities and the power associated with it.
20. Decision-making in the livestock management sector is rarely male-dominated.
21. Gender analysis informs the development of policies and programs that are gender-responsive.
22. Gender analysis aims to maintain existing gender disparities.
23. Gender analysis is an unnecessary process in animal husbandry.
24. Gender analysis considers the division of labour within households but not communities.
25. Gender analysis is primarily focused on quantitative data.
26. Gender analysis is essential for promoting gender inclusivity.
27. Gender analysis explores the relationship between gender norms and power relations.
28. Gender analysis only looks at gender roles in income-generating work.
29. Gender analysis can identify disparities in access to veterinary services in rural areas.
30. Gender analysis is mainly focused on the rights of men within a society.
31. The Harvard Analytical Framework examines gender roles and dynamics related to activities, resource allocation, and benefit control.
32. The 'Activity' profile in the Harvard Analytical Framework focuses on who has access to resources and controls their utilization.

33. The 'Access' and 'Control' profiles identify who holds access and control over resources, such as grazing land and veterinary supplies.
34. The Harvard Analytical Framework addresses gender inequalities directly and challenges traditional gender norms.
35. The Harvard Analytical Framework emphasizes the need for gender-specific divisions of labour within livestock farming activities.
36. The Moser Conceptual Framework identifies two types of gender needs: practical gender needs and strategic gender needs.
37. Practical gender needs involve activities that challenge existing gender relations and transform gender subordination.
38. The Moser Framework includes components such as gender role identification, gender needs assessment, and planning for balancing the triple role.
39. The Moser Framework is primarily focused on identifying practical gender needs, neglecting strategic gender needs.
40. Gender-sensitive policies address gender inequalities by promoting women's participation in decision-making and control over resources.
41. The Social Relations approach defines development as the enhancement of the well-being of men and women in society.
42. Social relations in the context of animal husbandry can lead to more equitable division of labor between men and women.
43. Institutional analysis in the Social Relations approach examines how local agricultural cooperatives influence gender dynamics.
44. Gender-blind policies challenge existing gender norms and seek to transform them.
45. Gender analysis matrices are primarily focused on labor allocation and time use within the livestock sector.
46. The 'Representation' step in the 4R Method involves surveying gender representation at all levels of decision-making in an organization.
47. The 'Resources' step examines the allocation of resources between women and men but does not consider access to time.
48. The 'Realia' step delves into the underlying reasons for gender disparities in representation and resource allocation.
49. The 'Realisation' step involves providing information about the current state of gender disparities but does not include recommendations.
50. The 4R Method is a comprehensive approach that systematically addresses gender disparities at all organizational levels.

51. Gender analysis encourages making assumptions about the lives of women and men, as it promotes a deeper understanding of their distinct needs.
52. Gender analysis aims to address challenges and disparities faced by one gender while neglecting the other.
53. Gender analysis assists in preventing gender-based discrimination, such as gender-based violence.
54. A comprehensive gender analysis helps in identifying the reasons behind gender disparities and challenging traditional gender norms.
55. Gender analysis leads to specific recommendations for addressing the needs of both women and men in an equitable manner.
56. Gender analysis in the livestock sector helps promote unequal access to resources and opportunities.
57. It's important to recognize that women often benefit the most from livestock extension services and technologies.
58. A gender analysis in the livestock sector mainly focuses on identifying disparities and challenges faced by women.
59. Gender analysis in the livestock sector contributes to enhancing gender responsiveness in policies and legislation.
60. By addressing gender disparities in the livestock sector, it's possible to create more equitable and sustainable development.
61. The Harvard Analytical Framework consists of the 'Activity' profile, 'Access' profile, and 'Control' profile to analyze gender dynamics.
62. Practical gender needs refer to the needs that have the potential to transform existing gender subordination.
63. The Moser framework focuses on gender role identification, gender needs assessment, and institutional analysis.
64. Gender-sensitive policies challenge existing gender norms and seek to achieve policy objectives.
65. The 4R method stands for Representation, Resources, Realia, and Realization.
66. Representation in the 4R method involves assessing gender representation at all levels of decision-making.
67. The 'Access' profile in the Harvard Analytical Framework analyzes who holds access to resources and benefits.
68. Gender-neutral policies consider men and women as equals but may not challenge existing gender norms.

69. The Moser framework includes planning for balancing the triple role of women, which includes reproductive, productive, and community management roles.
70. Gender-sensitive policies may address both practical and strategic gender needs.
71. The Social Relations approach primarily focuses on gender representation within organizations.
72. The 4R method is a four-step approach to assess gender disparities within organizations.
73. Gender analysis matrix framework examines the impact of gender on labor, time, resources, and sociocultural factors.
74. Practical gender needs are often related to day-to-day challenges and immediate requirements.
75. Gender analysis is mainly focused on promoting gender stereotypes.
76. Gender-sensitive policies assume men and women are the same and do not consider gender differences.
77. Representation in the 4R method examines the allocation of resources between men and women.
78. The Harvard Analytical Framework helps challenge and transform traditional gender norms.
79. Gender analysis can help organizations tailor their interventions to address the specific needs of different genders.
80. Strategic gender needs refer to the day-to-day needs of women and men in a community.
81. The Moser framework emphasizes the involvement of women and gender-aware organizations in the planning process.
82. Gender analysis primarily focuses on reinforcing traditional gender roles and stereotypes.
83. Gender analysis aims to identify and address gender disparities in various contexts.
84. Gender-positive policies focus on promoting gender stereotypes and gender biases.
85. The Social Relations approach is based on theoretical foundations like social relations and institutional analysis.
86. Gender-neutral policies challenge existing gender norms and roles.
87. Practical gender needs can include access to training and resources for specific tasks.

88. The 4R method helps organizations identify and remedy gender disparities.
89. Gender analysis is essential for understanding the distinct roles, challenges, and opportunities faced by women and men.
90. The Conceptual Framework for Gender Analysis in FSR/E aims to perpetuate the household-centric approach by focusing on male heads of households.
91. In the Benefits and Incentive Analysis, the term "control" refers to the freedom or permission to use a resource, while "access" refers to the power to decide how the resource is used.

Question 3. Fill in the Blanks

1. In various contexts, distinctions among women, men, girls, and boys become apparent in the distribution of ________, opportunities, restrictions, and power.
2. Gender analysis delves into how responsibilities are divided among different genders, how they access resources, exert control over them, and navigate societal ______________.
3. Gender analysis is crucial for gathering the necessary information to promote gender ______________.
4. Gender analysis encompasses both quantitative data and qualitative insights, considering aspects such as ______________, beliefs, attitudes, behaviors, and values.
5. Gender analysis examines how differences in societal roles and norms influence the lives of different __________.
6. Gender analysis seeks to answer questions such as who performs various roles, who has ownership of different resources, who holds authority and __________?
7. In the context of animal husbandry, a gender analysis helps identify who primarily makes decisions about livestock __________ and income use.
8. It is crucial to examine access to resources like land, credit, technology, and training to identify disparities based on __________.
9. In animal husbandry, women may have limited access to resources, such as land or training, due to legal or cultural __________.
10. Women often bear a significant burden in animal husbandry when it comes to caring for __________ animals.
11. Despite much of this burden being preventable through vaccination, women farmers rarely benefit from livestock __________.

12. A comprehensive gender analysis should encompass aspects like the political and legal __________.
13. Gender analysis assesses whether national policies, laws, and regulations acknowledge gender equality and __________ rights.
14. Gender analysis examines the division of tasks and responsibilities, including whether there is horizontal __________.
15. In productive work, women and men may be involved in income-generating activities, and gender analysis examines whether there is vertical __________.
16. In animal husbandry, decision-making processes may be male-dominated, and gender analysis can help identify who participates and at which __________.
17. Gender analysis helps in the development of policies, programs, and projects that are more equitable and responsive to the needs of both women and __________.
18. In a gender analysis, the division of labor in animal husbandry tasks is examined, including who is responsible for feeding, health management, and other related __________.
19. Gender analysis assesses whether there are disparities in access to resources like land, credit, technology, and __________.
20. Access to and control over resources, which encompass material and financial assets, are important aspects of gender analysis, and these resources can include intangible ones such as __________.
21. Gender analysis involves evaluating the availability of various services and institutions for women and men, considering whether there is equal access for both __________.
22. Gender analysis also examines how women and men are treated by service providers and whether there is any bias or discrimination in service __________.
23. A rigorous gender analysis ensures that policies and projects developed based on it have greater __________ among those affected by them.
24. In animal husbandry, women might have limited access to veterinary services, and gender analysis helps in understanding this gender-based difference and its underlying __________.
25. Gender analysis assists in designing programs that can provide women with access to veterinary services outside regular working hours, training opportunities in animal healthcare, and more equitable participation in __________ processes.

26. A comprehensive gender analysis examines the influence of gender roles and norms on the division of time between paid employment, unpaid work, and __________ activities.
27. Access to services and institutions is assessed in gender analysis, including the _________ of services available for both women and men.
28. Gender analysis helps in identifying and addressing disparities based on gender and promoting greater gender equality within the __________ sector.
29. Gender analysis informs the development of policies, programs, and projects that are more equitable, responsive, and __________ of both women and men.
30. The Harvard Analytical Framework consists of three interconnected components, focusing on understanding gender dynamics within activities, resource allocation, and control over benefits. These components include the 'Activity' Profile, 'Access' and 'Control' Profiles, and _________________ Factors.
31. The Moser conceptual framework for gender analysis and planning distinguishes between two types of gender needs: Practical Gender Needs and ________________________ Gender Needs.
32. The Social Relations approach to gender analysis emphasizes the importance of understanding ___________________ that perpetuate systemic disparities among various groups of people.
33. Gender-Positive Policies aim to challenge and transform gender norms, roles, and stereotypes to create more equitable opportunities and outcomes for all genders involved, while Gender-Neutral Policies assume that men and women are not equal but are the __________.
34. The 4R method involves four steps: Representation, Resources, Realia, and ________________.
35. Gender analysis is an essential tool for incorporating a gender perspective into policies, programs, and projects, helping to prevent making __________________ about the lives of women and men.
36. One of the practical benefits of gender analysis is that it reveals how cultural, economic, and legal factors disadvantage women and girls, as well as men and boys, by impacting their opportunities throughout their lives. It explores the connections between inequalities at various ___________________ levels.
37. Gender analysis in the livestock sector is essential for identifying disparities in roles, access to resources, and decision-making power

between men and women involved in livestock farming and ultimately promoting _______________ development in the sector.

38. Mechanization in the livestock sector can create additional burdens for women, such as increased labor demands for tasks like planting, weeding, and harvesting, primarily carried out by ____________.

39. In the context of gender analysis, addressing challenges faced by women in the livestock sector might involve providing training at convenient times, ensuring access to ____________, and promoting an equitable division of labor.

40. The Harvard Analytical Framework consists of three interconnected components: the 'Activity' Profile, the 'Access' Profile, and the '_________' Profile.

41. In the context of animal husbandry, the 'Activity' Profile helps identify individuals responsible for tasks like ______________, cleaning, milking, and administering medical care to the animals.

42. The 'Access' and 'Control' Profiles focus on the resources involved in these tasks, delineating who holds access to these resources and has control over their _______________.

43. Gender roles can play a significant role in determining access and control, with men often making decisions regarding land use and significant expenses, while women may oversee ______________.

44. The 'Influencing Factors' section of the Harvard Analytical Framework delves into the underlying reasons for gender disparities, which may include factors like _______________.

45. The Moser conceptual framework distinguishes between two types of gender needs: Practical Gender Needs and __________ Gender Needs.

46. In the context of animal husbandry, a practical gender need could involve access to training in modern milking techniques, while a strategic gender need might involve ______________.

47. The Moser framework includes several steps, including gender role identification, gender needs assessment, and ______________ within the household.

48. The framework emphasizes involving women and gender-aware organizations in the planning process to ensure that interventions meet their specific needs and ______________.

49. The Social Relations approach views development as the ongoing enhancement of ______________ well-being.

50. Social relations are defined as the structural connections that perpetuate systemic disparities among various ______________ of people.

51. In the livestock sector, social relations can lead to disparities in __________ and decision-making power due to different gender roles.
52. The Social Relations approach includes an institutional analysis, which examines how institutions can influence ______________ relations within the livestock sector.
53. Gender-blind policies often reinforce gender norms, roles, and stereotypes that perpetuate ______________ inequalities.
54. Gender-neutral policies operate within the existing gender division of resources and responsibilities without ______________ it.
55. Gender-sensitive policies presume men and women as equals and address gender norms, roles, and access to resources to achieve ______________ objectives.
56. Gender-positive policies actively work to change ______________ norms within a specific sector or activity.
57. Applying the gender analysis matrix framework involves looking at how gender differences impact labor, time, resources, and ______________ factors.
58. This framework helps stakeholders identify and address gender disparities, recognize the contributions of all genders, and work towards more ______________ practices.
59. The first step in the 4R method is 'Representation,' which involves surveying gender representation within an organization to provide a picture of the gender distribution at all levels of ______________.
60. In the livestock sector, 'Resources' in the 4R method examines the allocation of resources like money, time, and information between women and men and looks at how resources are ______________.
61. 'Realia' in the 4R method involves analyzing the underlying reasons for the gender distribution of representation and resource allocation, considering factors such as ______________.
62. The last step of the 4R method, 'Realisation,' focuses on formulating new objectives and measures to achieve ______________ in gender equality.
63. Gender analysis is crucial for uncovering the unique roles, challenges, and opportunities faced by men and women involved in ______________ and related activities.
64. Research has shown that women farmers often benefit the least from livestock extension services and technologies designed to improve production, creating additional burdens for ______________.

65. In the livestock sector, gender analysis helps develop strategies to overcome challenges faced by women, such as providing training at convenient times, ensuring access to credit, and promoting a _____________ division of labor.
66. Promoting equality in the livestock sector encourages initiatives that engage men in sharing household and childcare responsibilities, giving women more time to participate in livestock farming activities and _____________.
67. In conclusion, gender analysis in the livestock sector is essential for identifying and addressing gender disparities, advancing equitable development, and maximizing the sector's potential for all _____________ involved.
68. Activities Analysis indicates ________ does what in farming systems research. It helps screen the identification of problems, select research priorities, designate collaborating farmers, and design on-farm trials.
69. In the Resources Analysis stage, Table 2 outlines critical resources, including _______, _______, and _______. This analysis helps identify production constraints and proposed solutions.

ANSWERS

1. Multiple Choice Questions

1	b	To uncover and address gender-based disparities
2	c	It uncovers roles and responsibilities among women and men
3	b	Intangible resources like time and knowledge
4	b	To help enforce laws and regulations
5	b	Treatment by service providers
6	b	Selling livestock
7	b	Gender-based disparities in roles
8	b	It identifies challenges and disparities
9	c	By addressing gender disparities and promoting equality
10	c	Collection of gender- and sex-disaggregated data
11	c	Roles and responsibilities in livestock care
12	c	Resource distribution among different genders
13	c	It contributes to gender disparities
14	c	By establishing mobile veterinary clinics
15	c	To address gender disparities and promote equality
16	c	Unpaid care work

17	a	Harvard Analytical Framework
18	a	Practical gender needs address immediate challenges, while strategic gender needs focus on long-term transformation
19	b	Development
20	d	Gender-Positive Policies
21	a	Representation, Resources, Realia, Realization
22	a	Representation
23	a	Access and control of resources between women and men
24	c	Assessing the underlying reasons for gender distribution
25	b	Identifying challenges faced by women and developing strategies to overcome them
26	c	To uncover and address gender disparities and promote equitable development
27	b	Activity, access, and control
28	b	"Who does what?"
29	c	Practical and strategic gender needs
30	c	Formulating new objectives and measures to achieve gender equality
31	d	Gender-positive policies
32	c	Recognizing distinct needs, roles, and opportunities
33	b	Reduced opportunities for women's participation
34	c	Ensuring equitable access to resources
35	b	To challenge and transform gender norms
36	b	Adult women
37	a	Diagnosis
38	b	Interconnected farming systems
39	b	Who does what tasks in farming
40	d	Understanding resource access and control
41	c	Output of production
42	b	Persist with existing practices or adopt new ones
43	c	Ensuring fair and inclusive participation
44	c	Fair and inclusive participation
45	c	Planning and Design
46	b	Interconnected farming systems

47	d	Neglect of valuable contributions from other household members
48	c	To align technology with users' needs
49	c	Resource access and control
50	c	Activities map or profile
51	c	Understanding gender dynamics in livestock ownership
52	b	Household level production analysis worksheet
53	b	Division of labour within the household by economic level
54	c	Mapping keystakeholders and gender roles
55	b	Income levels
56	c	Division of labour within the household
57	c	Domestic & community work
58	b	By engaging a wide large of stakeholders
59	c	To address & mitigate obstacles
60	c	Gendered livestock stakeholder matrix
61	c	To understand diverse contributions
62	c	To promote long term empowerment
63	b	Current gender dynamics in animal husbandry
64	b	To track progress in promoting gender equity

Answer 2. True and False

S.No.	True or false	S.No.	True or false
1	True	2	True
3	False	4	False
5	True	6	False
7	False	8	False
9	True	10	False
11	False	12	True
13	False	14	False
15	True	16	False
17	True	18	True
19	True	20	False
21	True	22	False
23	False	24	False
25	False	26	True

27	True	28	False
29	True	30	False
31	True	32	False
33	True	34	False
35	True	36	True
37	False	38	True
39	False	40	True
41	True	42	True
43	True	44	False
45	True	46	True
47	False	48	True
49	False	50	True
51	False	52	False
53	True	54	True
55	True	56	False
57	False	58	False
59	True	60	True
61	True	62	False
63	True	64	True
65	True	66	True
68	True	68	True
69	True	70	True
71	False	72	True
73	True	74	True
75	False	76	False
77	False	78	False
79	True	80	False
81	True	82	False
83	True	84	False
85	True	86	False
87	True	88	True
89	True	90	False
91	False		

Answer 3. Fill in the Blanks

1	Resources
2	Interactions
3	Inclusivity
4	Preferences
5	Genders
6	Who participate in decision making
7	Management
8	Gender
9	Norms
10	Diseased
11	Vaccines
12	Framework
13	Women's
14	Segregation
15	Segregation
16	Levels
17	Men
18	Tasks
19	Training
20	Knowledge
21	Genders
22	Provision
23	Credibility
24	Causes
25	Decision-making
26	Volunteer
27	Availability
28	Livestock
29	Inclusive
30	Influencing Factors
31	Strategic
32	Social Relations
33	Same
34	Realisation
35	Assumptions

36	Societal
37	Equitable
38	Women
39	Credit
40	Control
41	Feeding
42	Utilization
43	Day-to-day expenses and tasks
44	Cultural norms and expectations
45	Strategic
46	Challenging traditional gender roles
47	Disaggregating control of resources and decision-making
48	Challenges
49	Human
50	Groups
51	Workload
52	Gender
53	Gender
54	Challenging
55	Policy
56	Gender
57	Sociocultural
58	Inclusive
59	Decision-making
60	Allocated
61	Cultural norms and biases
62	Gender
63	Animal husbandry
64	Women
65	Equitable
66	Training
67	Individuals
68	Who
69	Land, Water, Labour

5

Gender Mainstreaming in Animal Husbandry

Chapter Overview

This chapter delves into the concept and application of gender mainstreaming within the context of animal husbandry and the livestock sector. Gender mainstreaming is a comprehensive strategy aimed at promoting gender equality by integrating a gender perspective into all aspects of policies, programs, and projects. It recognizes and addresses the distinct roles, needs, and contributions of women and men involved in animal husbandry. This chapter provides insights into why gender mainstreaming is essential in this sector and outlines practical approaches for its implementation.

SECTION A: THEORY

Gender Mainstreaming

The concept of gender mainstreaming emerged as a response to the frustration with strategies devised during the 1970s and 1980s to integrate women into development efforts, strategies that had shown limited progress in promoting gender equality. The term "gender mainstreaming" was introduced after the third world conference on women in Nairobi in 1985 and gained explicit recognition and adoption during the Fourth World Conference on Women in Beijing in 1995. This adoption was prompted by the realization that various policies, programs, and actions were achieving minimal impact in terms of advancing gender equality within societies. Subsequently, numerous international and national organizations embraced gender mainstreaming strategies.

Gender mainstreaming involves the integration of a gender equality perspective into every stage and aspect of policies, programs, and projects. Recognizing that women and men have distinct needs and live under varying conditions and circumstances, often facing unequal access to power, resources, human rights, and institutions, including the justice system, gender mainstreaming seeks to account for these differences when crafting, executing, and assessing policies, programs, and projects. The primary objective is to ensure that these initiatives benefit both women and men without exacerbating inequalities but rather promoting gender equality. Essentially, gender mainstreaming serves as a tool for addressing and rectifying often subtle gender disparities, contributing to the overall goal of achieving gender equality.

Definitions of Gender Mainstreaming

It is the process of assessing the implications for women and men of any planned action, including legislation, policies or program, in any area and at all levels. It is a strategy for making the concerns and experiences of women as well as of men an integral part of the design, implementation, monitoring and evaluation of policies and program in all political, economic and societal spheres, so that women and men benefit equally, and inequality is not perpetuated. The goal of mainstreaming is to achieve gender equality.

United Nations Economic and Social Council (ECOSOC), 1997

"Gender mainstreaming is the integration of the gender perspective into every stage of policy processes design, implementation, monitoring and evaluation – with a view to promoting equality between women and men and combating discrimination.

(European Institute of Gender Equality)

It means assessing how policies impact on the life and position of both women and men – and taking responsibility to re-address them if necessary. This is the way to make gender equality a concrete reality in the lives of women and men creating space for everyone within the organizations as well as in communities - to contribute to the process of articulating a shared vision of sustainable human development and translating it into reality.

(The European Commission's Directorate-General for Employment, Social Affairs and Equal Opportunities)

Gender Mainstreaming Efforts few real-world examples

Women's Involvement in Dairy Cooperatives (India)

In India, organizations such as the National Dairy Development Board (NDDB) have implemented programs to promote gender mainstreaming in the dairy sector. One notable initiative is the involvement of women in dairy cooperatives. Through programs like Operation Flood, women are encouraged to participate in dairy farming activities, including animal rearing, milk production, and cooperative management. By integrating women into dairy cooperatives, these programs empower women economically, enhance household nutrition, and contribute to poverty reduction in rural areas.

Gender-Sensitive Poultry Extension Services (Africa)

In many African countries, organizations like the International Livestock Research Institute (ILRI) work to mainstream gender considerations in poultry production and extension services. These efforts include providing gender-sensitive training and support to both men and women involved in poultry farming. For instance, extension agents are trained to recognize and address gender-specific constraints faced by women poultry farmers, such as access to resources, training, and markets. By incorporating gender perspectives into poultry extension services, these programs aim to improve productivity, income, and food security for households.

Women's Participation in Animal Husbandry (Latin America)

In countries like Colombia and Guatemala, projects supported by organizations such as Heifer International focus on promoting women's participation in animal husbandry and livestock management. These projects provide women with training, resources, and support to engage in various aspects of animal husbandry, including feeding, breeding, and healthcare. By empowering women as active participants in animal husbandry, these initiatives enhance household food security, increase income-generation opportunities, and promote gender equality within rural communities.

In each of these examples, gender mainstreaming efforts in animal husbandry, dairy, and poultry sectors aim to address the specific needs, priorities, and constraints faced by both men and women, ultimately contributing to more inclusive and sustainable agricultural development.

Sources

https://globalagriculturalproductivity.org/improving-gender-and-nutrition-outcomes-of-women-poultry-farmers/

https://tanagerintl.org/portfolio/improving-gender-and-nutrition-with-stronger-poultry-markets/

https://www.heifer.org/our-work/flagship-projects/rural-entrepreneurs-project-mexico.html

What is mainstreaming?

To understand gender mainstreaming, it is critical we understand the mainstream. Mainstreaming refers to a strategy utilized in development policy and practice today for increasing attention and resources to a wide range of specific development issues which might otherwise be neglected -- including the environment, human rights, disability, and the situation of children. Gender mainstreaming refers to the use of the mainstreaming strategy to bring attention and resources to critical issues of gender equality, women's rights, and empowerment, which otherwise can be ignored or neglected

Concept of Gender Mainstreaming

Gender mainstreaming is a comprehensive approach aimed at incorporating the perspectives, experiences, knowledge, priorities, and interests of both women and men into all aspects of policymaking, planning, implementation, and monitoring across social, political, and economic actions. This involves various levels of change, from internal organizational adjustments to more profound structural transformations to foster a more equitable work environment. As organizations engage in project implementation, multiple changes are necessary to ensure gender mainstreaming at all stages, including design, implementation, monitoring, and evaluation.

These changes may involve influencing project goals, strategies, and resource allocations from the outset and providing specialized inputs like gender analysis and technical assistance at different project stages. Gender mainstreaming is a dynamic process that acknowledges the importance of actively considering gender dynamics and inequalities throughout the policy and project cycle to promote gender equality and inclusivity. It underscores the significance of recognizing and addressing the unique needs and experiences of women and men in all aspects of social, political, and economic actions.

Evolution of Gender Mainstreaming Approaches

The evolution of gender mainstreaming approaches started with the welfare approach and progressed through the Women in Development (WID), Gender Efficiency and Empowerment approaches, leading to the current Gender Mainstreaming approach, which aims to integrate gender considerations into all aspects of policy and program development for the promotion of gender equality.

a) Welfare Approach

In the early 1970s, development policies primarily addressed the needs of impoverished women within the context of their roles as wives and mothers. The welfare approach concentrated on improving maternal and child health, child care, and nutrition. This approach operated under the assumption that the benefits of macroeconomic strategies focused on modernization and growth would eventually benefit the poor, including poor women, as their husbands' economic positions improved.

For example, in a developing country, a welfare approach might involve government initiatives focused on improving maternal and child health. For instance, they could establish health clinics that provide free prenatal and postnatal care to pregnant women and healthcare services to children. The assumption is that by enhancing the health of mothers and children, it will indirectly benefit the family's overall well-being as fathers become more economically productive.

However, Danish economist Easter Boserup(1970) challenged these assumptions systematically in her book 'Women's Role in Economic Development.' She argued that, contrary to expectations, women were not experiencing improved status as men's economic situations improved. Instead, women were increasingly associated with traditional and backward roles, while men were linked with modern and progressive roles.

b) Women in Development Approach (WID)

In response to growing research and activism regarding women's status and the women's movement in the USA and Europe, the United Nations declared 1975 the International Year for Women, followed by the International Women's Decade from 1975 to 1985. This marked a significant shift in recognizing and addressing women's needs and concerns within the development sector. The WID approach emerged in response to the realization that women's concerns had been neglected in economic development processes. WID aimed to integrate women into the development process, focusing on their productive roles, leading to the creation of numerous women's income-generation programs.

For example suppose a development organization adopts the WID approach. In an agricultural community, they may introduce training programs that teach women modern farming techniques and provide them with access to resources like better seeds and tools. This empowers women to actively contribute to the community's agricultural productivity, going beyond their traditional roles, which might have been limited to household chores.

c) Gender Efficiency and Gender Empowerment Approaches

By the mid-1990s, the gender approach gained widespread adoption by governments, donor organizations, and NGOs. Gender efficiency analysis played a pivotal role in bringing women's concerns and gender differences to the forefront of development. This approach argued that understanding men's and women's roles and responsibilities in all development interventions improved project effectiveness and ensured the active participation of both women and men in development. The gender empowerment approach aimed at empowering women as agents of change, working with them at the community level to build organizational skills and self-esteem through active involvement in identifying needs and managing change.

For example in a rural village, a development project takes a gender-efficient approach. They conduct a comprehensive analysis of the roles and responsibilities of both men and women in agriculture. As a result, they design interventions that ensure both genders benefit from agricultural improvements. For instance, they might provide training on sustainable farming techniques and involve women in decision-making regarding crop selection, leading to increased yields and improved livelihoods for both women and men.

Likewise, a gender empowerment project could work with women in the same village. They might establish women's self-help groups that not only focus on income-generating activities but also on building leadership and organizational skills. These groups enable women to actively participate in community decisions, challenging traditional norms and empowering women to influence change in various aspects of their lives.

Critics of the Gender Efficiency approach argued that it succeeded in elevating women's concerns in the development process but often focused on what women could do for development rather than what they could do for themselves. The Gender Empowerment approach, while creating opportunities, was at times misinterpreted as an end rather than a means to an end.

d) Gender Mainstreaming Approach

Following the Beijing Platform for Action in 1995, the concept of gender mainstreaming gained widespread acceptance. Gender mainstreaming is a commitment to ensuring that the concerns and experiences of both women and men are integral to the design, implementation, monitoring, and evaluation of all legislation, programs, and policies to promote gender equality.

Gender mainstreaming does not exclude projects targeting women; instead, it considers women as a specific target group with gender equality as the goal. It supports and promotes projects designed as strategic interventions to address gender inequality and promote social equality. For example imagine a national government embracing gender mainstreaming. In education policy, they ensure that both girls and boys have equal access to quality education. They monitor school curricula to eliminate gender biases and stereotypes, making sure that textbooks and teaching materials promote gender equality. By doing so, they aim to create an inclusive educational system where both genders have the same opportunities and outcomes.

Likewise in a healthcare context, a gender mainstreaming approach could involve designing and implementing a national health insurance system that considers the unique healthcare needs of women and men. It would provide comprehensive coverage for issues such as maternal health and family planning while also addressing men's health concerns. This approach promotes gender equality by addressing the specific health requirements of each gender.

In each of these approaches, the evolution from the welfare approach to gender mainstreaming reflects a growing understanding of the complexities of gender disparities and the need for more comprehensive and equitable solutions to address them. Gender mainstreaming, in particular, seeks to integrate gender equality into all aspects of development, ensuring that both women and men benefit from these initiatives and contribute to their success.

Central principles of gender mainstreaming are:

1. *Gender equality is the goal of gender mainstreaming.*

   Gender mainstreaming has limited value unless it is explicitly focused on promoting and monitoring positive gender equality results in all development interventions.

2. *Gender mainstreaming is relevant for, and should be utilized in, all sectors and policy areas.*

   Gender mainstreaming is important at both national and sub national intervention levels. It is important at the national policy level, in goal setting, developing overarching strategic frameworks, national development plans and programming approaches, and in creating and following up on budgets. At sub national level, it is critical in the design, implementation and follow-up of programmes, projects, and services.

3. *Gender mainstreaming involves both fully integrated and targeted actions for achieving gender equality results– the 'twin-track' approach.*

   Gender mainstreaming should both "integrate" attention to gender equality in routine processes and procedures and employ "targeted interventions" to address specific constraints and challenges faced by women or men and girls or boys. Targeted activities through gender mainstreaming aim to empower women and girls, through for example, providing information, developing skills, increasing social capital and enhancing self-esteem. Some common targeted activities include literacy training; programmes to increase women's access to income; providing access to natural and productive resources, services and institutions; legal literacy programmes; support to networking; and providing safe women-only spaces for mobilization and empowerment activities. The integrated approach in gender mainstreaming aims to make all existing and planned policies, programmes, activities, outcomes, and results gender-responsive by systematically and effectively identifying and taking explicit action to address relevant gender-equality dimensions as an essential part of all development policy and practice, with the overall explicit goal of promoting gender equality and achieving positive gender equality results.

Characteristic of gender mainstreaming

i. Gender mainstreaming is a globally accepted strategy to promote gender equality. It's not a final goal on its own, but rather a method, an approach to achieving the ultimate objective of gender equality.

ii. Gender mainstreaming acknowledges the interconnected and complementary roles of both men and women. Changing one role would naturally impact the other, aiming to reshape not just the unequal interactions between genders but also the systems generating such inequality.

iii. Gender-related matters are not limited to a specific area; they demand comprehensive attention. Gender concerns should be an integral part of regular institutional activities, not solely relegated to specialized women's organizations. The responsibility of addressing this lies with the entire institution rather than any single person or sector.

iv. Gender perspectives and the aspiration for gender equality are fundamental to all undertakings. They must be incorporated into policy formulation, research, advocacy, discussions, legal frameworks, resource allotment, and the entire cycle of program and project development, execution, and evaluation.

Blueprint for gender mainstreaming

A standardized formula or blueprint for implementing gender mainstreaming does not exist, as the approach varies depending on the context. The strategy of mainstreaming is adapted differently across various activities like research, policy development, policy analysis, program delivery, and technical assistance. Each area of work has distinct opportunities and processes associated with it.In practical terms, the gender mainstreaming methodology involves applying a gender-focused perspective to every action, intervention, policy, program, or project intended to be undertaken. The following initiatives are part of gender mainstreaming practices:

1. **Gender Analysis:** Conducting a gender analysis to identify existing inequalities between men and women that require addressing. In livestock production, this could involve assessing the division of labor, resource access, and decision-making within farming households. For example, a gender analysis might reveal that women are responsible for small ruminant care, while men manage larger livestock.
2. **Equal Opportunities:** Ensuring equal opportunities for all individuals and implementing gender-specific measures where noticeable inequalities persist. In livestock farming, this means providing women with access to resources like land, credit, and training. For instance, in

many regions, women may be excluded from owning land and availing loans so a gender mainstreaming initiative could work to change policies and practices to ensure women can own land and access credit to purchase livestock.

3. **Participatory Involvement:** Striving for equitable participation of both men and women in determining priorities and contributing to the design, development, implementation, direction, and monitoring of programs. This often includes giving girls and women a platform to voice their opinions. In livestock production, this could involve including women in discussions on breed selection, animal health practices, and marketing strategies. For example, women may have insights into the specific needs of small ruminants and can contribute to decisions about their care
4. **Gender Budgeting:** Incorporating gender budgeting practices to evaluate the gender-specific impacts of financial allocations. In livestock production, it could involve analysing how investments in animal healthcare, feed, and infrastructure affect gender roles. For instance, budget analysis might reveal that women primarily handle animal healthcare, and investments should be made to improve their access to veterinary services and knowledge.
5. Providing organizational training on tools that facilitate an environment for initiating institutional changes. For example, training programs can educate livestock extension workers on the specific needs and challenges faced by women in livestock farming. They can then adapt their advice and services accordingly. It is worth training women on value addition of livestock products which help in their empowerment through increased income through sale of value-added products like paneer, khoa etc.
6. Carrying out participatory gender audits to comprehensively assess gender-related aspects. Conducting gender audits in livestock production systems might uncover disparities in income generation, decision-making power, or access to extension services. For instance, a gender audit might reveal that women are not accessing extension services, and strategies could be developed to make these services more accessible and tailored to their needs.

To sum up, while a universal blueprint for gender mainstreaming doesn't exist, the approach involves undertaking every activity with a gender-aware perspective and employing various initiatives such as gender analysis, equal opportunities, participatory involvement, and gender audits to promote gender equality.

Gender mainstreaming in practice

The following three aspects need to be kept in mind when we want to apply the gender mainstreaming strategy

1. **Technical Dimension:**

   To effectively carry out gender mainstreaming, it is crucial to have access to accurate and reliable data. This data should be sex-disaggregated and collected from a substantial sample size. Furthermore, a robust theoretical foundation is essential for understanding the intricacies of gender dynamics. Expertise from gender specialists is also indispensable, as they can identify opportunities and interpret the requirements for achieving gender equality across various groups.

2. **Political Dimension:**

   Gender mainstreaming goes beyond merely increasing women's participation within a development framework that may be unjust and flawed. It also involves advocating for women's elevation to decision-making positions. This means actively supporting women's collaborative efforts to redefine the priorities of development agendas. The ultimate goal is to create an empowering environment where women can assess their circumstances and voice their preferences and concerns. This political dimension of gender mainstreaming is essential for effecting real change. Now in local bodies like Panchayats and Municipalities one third seats are reserved for women to voice their concerns. Surprisingly 88 % of the 12 million SHGs in India are all women members.

3. **Cultural Dimension:**

   In any initiative, it's crucial to acknowledge that it operates within a specific cultural context. The execution and interpretation of gender mainstreaming initiatives are significantly influenced by socio-cultural norms, prevailing thoughts, and ideologies at the local level. These factors exert considerable influence on the effectiveness of any project. Therefore, it is imperative to take them into account when implementing the mainstreaming strategy. Recognizing the cultural dimension is essential for ensuring that the strategy aligns with the local context and is sensitive to the diverse cultural landscapes in which it operates.

   Only by addressing these three key aspects can we truly make gender mainstreaming a reality in our development initiatives.

Approaches to Gender Mainstreaming

Gender mainstreaming is an approach that encompasses both the planning of activities and the transformation of institutions in alignment with global objectives concerning gender equality and the empowerment of women. To effectively implement gender mainstreaming, there is a need to systematically incorporate gender perspectives into policies, programs, and thematic areas.

A. Gender mainstreaming at programmatic level
B. Incorporating Gender Mainstreaming into Institutions

A. Gender mainstreaming at the programmatic level

i. Gender Analysis
ii. Programme Design
iii. Human and Financial Resource Allocation
iv. Implementation
v. Monitoring and Evaluation

Step i. Gender Analysis

Conduct a comprehensive examination of the context through gender analysis. Determine how and why certain issues impact women and men differently within a specific setting or developmental sector. Identify potential solutions to address these disparities.

For example before implementing a project in the goat and dairy sector, we conduct a gender analysis to understand how men and women are involved and impacted differently in goat and dairy farming. Identify specific gender-related challenges and opportunities. For example, the analysis might reveal that women are primarily responsible for attending to goats while kidding and taking care of the young ones. Gender analysis also shows that women have limited access to training and veterinary services on management of kids.

Step ii. Programme Design

Utilize the insights gained from gender analysis to shape program design. This involves selecting key priorities, target demographics, and coverage areas. Integrate these aspects into program outcomes, indicators, and methods of intervention.

For example, based on the findings, design a program that aims to empower women in management of kids to reduce kid mortality and increase their body weight by providing them with a tailor-made training on these aspects. Include specific indicators in the program design to measure increased participation and decrease in kind mortality and increase in body weight of kids.

Step iii: Resource Allocation

Ensure that sufficient resources are allocated to effectively tackle gender equality considerations throughout the program cycle.

This might involve dedicating adequate financial resources to training programs for women in goat management and providing them with access to veterinary services.

Step iv: Programme Implementation

Promote collaborations across various sectors and disciplines by involving diverse stakeholders, including women's organizations, in the execution of programs.

For example, during the implementation phase, involve diverse stakeholders, including women's organizations, in executing the program. Collaborate with these organizations to organize training sessions and workshops focused on enhancing women's skills and knowledge in goat farming.

Step v: Monitoring and Evaluation

Employ robust monitoring and evaluation techniques to establish a factual foundation for strategic decisions related to gender equality. This also facilitates enhanced development planning and holds institutions accountable for their commitments to gender equality.

For example, in goat farming we eemploy robust monitoring and evaluation techniques to assess the program's impact on gender equality in goat farming Collect data on women's increased participation in all aspects of goat management including decision-making and any improvements in their economic empowerment through goat farming. Use this data to make informed decisions about program adjustments and to be accountable for the program's commitment to gender equality in the sector.

B. Incorporating Gender Mainstreaming into Institutions

i. Establishment of institutional Structures:

Create institutional mechanisms like gender units and gender focal point systems to support the integration of gender mainstreaming.

For example, create a Gender and Livestock Unit within an agricultural research institution focused on animal husbandry. This unit could be responsible for conducting gender-sensitive research on livestock management, providing training to livestock extension workers with a gender perspective, and ensuring that gender-related data is collected and analysed. Employing more women extension personnel in livestock research and development institutions facilitate working more closely with women groups like WSHGs, all women Dairy Coops etc.

ii. Resource allocation:

Devote adequate financial and human resources to support endeavours aimed at incorporating gender considerations. For example, allocate sufficient financial and human resources to support gender mainstreaming initiatives within animal husbandry. Adequate resources are essential for conducting gender analysis, developing gender-responsive programs, and providing training and capacity-building activities. Allocate a specific budget for gender-focused training programs for animal husbandry professionals and farmers This budget could cover the costs of designing training materials, organizing workshops, and hiring gender experts to facilitate the sessions in addition to assess the impact of these extension initiatives on their family income.

iii. Implementation of Accountability Mechanisms:

Introduce mechanisms that ensure accountability for the advancement of gender equality within the institution. For example, develop a set of gender-sensitive indicators to assess the impact of livestock development projects. These indicators could measure improvements in women's participation in decision-making processes related to animal husbandry, changes in access to livestock-related resources for women, and the integration of women's knowledge in animal health and nutrition practices.

Need for Gender Mainstreaming

In the context of India, gender disparities persist in various aspects of society, reflecting a deeply entrenched inequality between men and women. The following points shed light on the multifaceted nature of this gender inequality:

Gender Inequality

i. **Imbalanced Sex Ratio**: India faces a stark gender imbalance in rural areas, with 985 women for every 1000 men in 2023 compared to 943 in 2011. Many women never get the chance to be born, and others succumb to easily preventable diseases during their lifetimes. Despite Govt. restrictions female foeticides are being reported mainly because of the desire to have a male child.

ii. **Nutritional Stress**: A significant portion of Indian women experiences chronic nutritional stress, leading to conditions like anemia and malnourishment. Nutritional discrimination within families often results in women receiving less food than they need. Culturally women are the last to take food and there is a bias in providing food first to boys and then to girls in many families especially in rural areas of India.

iii. **Limited Educational Opportunities**: Gender disparities extend to education, where the female literacy rate is 70.3 % when compared to 84.7% for males. Girls face challenges in accessing and completing their education, with many dropping out to work in their families' homes or fields.

iv. **Undervalued and Unrecognized Labor**: Women's work, both within households and communities, is often unpaid and goes unnoticed. Women work longer hours than men, bearing the major share of the household and community workload.

v. **Underrepresentation in the Workforce**: There are far fewer women in the paid workforce compared to men, and more women experience unemployment. Those who do work earn significantly lower wages than men for similar roles.

vi. **Informal Sector Vulnerability**: Women predominantly work in the informal sector, characterized by lower wages and lack of labour law protections. They are often relegated to lower-paid positions within organizations and may engage in piecework or subcontracting at exploitative rates. For example, in manual ploughing operations in 2009, where men were paid 103 per day, while women were paid 55, a wage gap ratio of 1.87. According to the World Inequality Report 2022 estimates, men earn 82 per cent of the labour income in India, whereas women earn 18 per cent.

vii. **Underrepresentation in Governance**: Women have limited representation in governance and decision-making roles. women represent only 15.1% of parliamentarians. As per Indian Administrative Services (IAS) data and the central government's employment census of 2011, less than 11 per cent of its total employees were women. In 2020, this reached 13 per cent

viii. **Legal Discrimination in Property Rights**: Despite constitutional guarantees, women face legal discrimination in terms of land and property rights. Most women do not own property in their names or receive a share of parental property.

ix. **Violence against Women**: Women face violence throughout their lives, both within and outside their families. One study found that 4 out of 10 women in India have experienced domestic violence in their lifetime and 3 out of 10 women experienced domestic violence in the past year. Most crimes against women often go unreported.

x. **Alarming Crime Statistics**: Data collected by women's groups reveals alarming statistics, including molestation every 26 minutes, a rape every 34 minutes, sexual harassment every 42 minutes, kidnapping every 43 minutes, and a woman being killed every 93 minutes. Many of these crimes are committed by individuals known to the victims.

xi. **Cultural and Organizational Vulnerability**: The vulnerability of women in the Indian context is predominantly a result of cultural and organizational factors rather than biological or physiological ones. Understanding and addressing these cultural and organizational aspects is essential to combat gender inequality effectively.

Gender mainstreaming is a crucial concept and strategy aimed at promoting gender equality and addressing gender-based disparities in all aspects of society. It involves incorporating a gender perspective into the design, implementation, monitoring, and evaluation of policies, programs, and projects across various sectors. The need for gender mainstreaming arises from several important reasons:

i. **Promotion of Gender Equality**: Gender mainstreaming seeks to challenge and transform traditional gender roles and norms, creating a more equitable society where individuals of all genders have equal rights, opportunities, and access to resources.

ii. **Addressing Gender Disparities**: In many societies, gender disparities are prevalent across various domains, including education, healthcare, employment, and political representation. These disparities often result in unequal opportunities and outcomes for different genders. Gender mainstreaming serves as a crucial approach to not only recognize but also address these imbalances through targeted and strategic interventions.

iii. **Effective Policies and Programs**: By incorporating a gender perspective, policies and programs can be better tailored to meet the specific needs and concerns of different genders, ensuring they are more effective and relevant.

iv. **Enhancing Social and Economic Development**: Gender equality has been linked to increased economic growth and improved social indicators. Gender mainstreaming can contribute to the overall development of a society by utilizing the skills and talents of all individuals.

v. **Reducing Gender-Based Violence**: Addressing unequal power dynamics and traditional gender norms can contribute to the reduction of gender-based violence, as well as challenging harmful stereotypes and attitudes.

vi. **Ensuring Human Rights**: Gender mainstreaming is aligned with the principles of human rights, ensuring that everyone is treated with dignity and respect regardless of their gender identity.

vii. **Fulfilling International Commitments**: Many countries have committed to international agreements and conventions related to gender equality. Gender mainstreaming is often a requirement to fulfil these commitments.

viii. **Fostering Inclusive Governance**: Incorporating diverse voices and perspectives into decision-making processes contributes to more inclusive and representative governance.

ix. **Improving Health Outcomes**: Gender mainstreaming can lead to better health outcomes by considering the unique health needs and vulnerabilities of different genders.

x. **Empowering Marginalized Groups**: Gender mainstreaming can particularly benefit marginalized groups such as women and girls, transgender individuals, and non-binary people, by providing them with equal opportunities and recognition.

xi. **Breaking Stereotypes**: Challenging traditional gender stereotypes and roles can lead to more open-minded and inclusive societies that value individuality and diversity.

xii. **Long-Term Social Change**: Gender mainstreaming aims for lasting social change by embedding gender equality into the very fabric of society, thereby contributing to more sustainable and equitable societies.

In conclusion, gender mainstreaming is a crucial strategy to promote gender equality, challenge gender-based disparities, and create a more inclusive and just society. It recognizes that gender is a significant factor that shapes people's experiences and opportunities and aims to rectify historical and structural inequalities

Need for Gender mainstreaming in the livestock sector

Gender mainstreaming in the livestock sector is essential for several reasons, as it recognizes the different roles, needs, and contributions of men and women in livestock production activities. Here are some key reasons why gender mainstreaming is important in the livestock sector:

i. **Equity and Social Justice**: Gender mainstreaming promotes equality between men and women by acknowledging and addressing the existing gender disparities in the sector. Women often play significant roles in livestock-related tasks, but their contributions are often undervalued and overlooked. Mainstreaming gender helps ensure that both men and women have equal access to resources, benefits, and opportunities.

ii. **Enhanced Productivity**: Women constitute a substantial portion of the agricultural labour force, including the livestock sector. Integrating gender perspectives can lead to increased productivity as women's contributions are acknowledged and supported. When women have access to resources like training, credit, and technology, livestock productivity can improve.

iii. **Improved Livelihoods:** Gender mainstreaming can positively impact household livelihoods by recognizing and valuing women's roles as producers, caregivers, and income earners. Improving women's access to resources and decision-making processes can enhance their capacity to generate income and contribute to the overall well-being of their families.

iv. **Resilience and Food Security**: In many regions, women are responsible for managing small-scale livestock production. When women's knowledge and expertise are integrated into decision-making processes, it can lead to more effective and sustainable livestock production practices. This, in turn, contributes to food security and resilience in the face of climate change and other challenges.

v. **Inclusive Policies and Programs**: Gender mainstreaming helps policymakers and development organizations design more effective and targeted policies and programs. When gender-specific needs and priorities are considered, interventions can be tailored to address the diverse requirements of men and women in the livestock sector.

vi. **Empowerment and Participation**: Integrating gender perspectives empowers women to become active participants in decision-making processes related to livestock management, technology adoption, and resource allocation. This empowerment contributes to greater agency and autonomy for women within their households and communities.

vii. **Knowledge Sharing**: Women often possess unique knowledge and skills related to animal husbandry, animal health, and traditional practices. Gender mainstreaming encourages the sharing of this valuable knowledge, contributing to a more holistic understanding of livestock management.

viii. **Sustainable Development**: Gender equality is a fundamental principle of sustainable development. By addressing gender disparities and promoting women's involvement in the livestock sector, societies can move closer to achieving the Sustainable Development Goals (SDGs), particularly those related to poverty reduction, gender equality, and sustainable agriculture.

In conclusion, gender mainstreaming in the livestock sector is crucial for achieving equitable and sustainable development. It recognizes the roles and contributions of both men and women, and it leads to more effective and efficient livestock management practices that benefit communities, households, and individuals alike.

Summary

- Gender mainstreaming is a strategy for promoting gender equality in all aspects of policy, program, and project development.
- It emerged as a response to the limitations of previous efforts to integrate women into development in the 1970s and 1980s.
- Gender mainstreaming was formally recognized at the Fourth World Conference on Women in Beijing in 1995.
- Its central principle is achieving gender equality in all development interventions.
- It involves both fully integrated and targeted actions, often called the "twin-track" approach.
- Gender mainstreaming is relevant in all sectors and at various levels of intervention.
- It encompasses gender analysis, program design, resource allocation, implementation, and monitoring and evaluation.
- Gender mainstreaming in the livestock sector in India focuses on women's participation, resource access, capacity building, income generation, and gender-responsive research.
- Implementing gender mainstreaming can lead to more equitable and sustainable development in the livestock sector.

SECTION B : DESCRIPTIVE QUESTIONS

1. Elaborate an example to illustrate gender mainstreaming in the delivery of veterinary services?

Traditional Approach (Without Gender Mainstreaming):

In a traditional approach to deliver veterinary services, the focus might solely be on providing medical care to animals without considering the broader social context. Services might be planned and implemented without considering the different roles and responsibilities of women and men in animal husbandry.

Gender Mainstreaming Approach:

Applying gender mainstreaming to the delivery of veterinary services would involve considering the specific needs and contributions of both women and men involved in animal husbandry. Here's how it might work:

i) Needs Assessment: Before designing veterinary service programs, a comprehensive needs assessment would be conducted, considering the roles of both women and men in animal care. It would involve understanding how animals are cared for, who is responsible for what tasks, and any barriers or challenges faced by different genders.

ii) *Design and Implementation*: Based on the needs assessment, veterinary services would be designed to address the diverse needs of women and men. For example, if women are primarily responsible for small animal care (e.g., poultry, goats), services could be *scheduled at times that are convenient for them*, and training materials could be provided in *formats accessible to them.*

iii) *Access and Outreach:* Gender mainstreaming would also consider the accessibility of veterinary services. If women have *limited mobility due to cultural norms, efforts could be made to provide services closer to their homes or in a way that respects their privacy.*

iv) *Training and Capacity Building:* Training sessions related to animal health and husbandry would be designed in a way that accommodates the learning preferences and availability of both women and men. This might involve offering separate sessions, if needed, or creating an inclusive learning environment.

v) *Monitoring and Evaluation:* Gender-sensitive indicators would be established to measure the impact of veterinary services on both women and men. For instance, data would be collected on the changes in animal health outcomes, as well as any changes in the division of labour and decision-making within households related to animal care.

By applying gender mainstreaming to the delivery of veterinary services, the aim is to ensure that both women and men benefit equally from the services provided. This approach recognizes that gender norms and roles can influence how services are accessed and utilized and seeks to promote more equitable outcomes while avoiding the perpetuation of existing gender inequalities.

2. Discuss how the concept of gender mainstreaming can be applied in the context of animal husbandry and the livestock sector in India?

Gender mainstreaming can be applied in the context of animal husbandry and the livestock sector in India in following ways

1. *Staffing and Personnel Policies*: To promote gender mainstreaming, a livestock organization in India may implement policies that aim to hire more women in various roles, such as veterinary technicians, livestock extension officers, or administrative positions. They may also set targets for having a certain percentage of leadership positions held by women. For instance, they could strive for 30% of managerial roles to be occupied by women within a specified timeframe.
2. *Structural Changes for an Egalitarian Work Culture/Promoting an Egalitarian Work Culture*: The organization can take steps to foster

an egalitarian work culture by conducting gender sensitivity training for all staff. This training would address biases and stereotypes, raise awareness of gender-related issues, and encourage respectful and equal treatment of male and female employees. Additionally, the organization could establish mentorship programs for women in the livestock sector to provide them with support and guidance for career development.

3. *Mainstreaming Gender into Project Design*: When planning a project related to livestock development, the organization would ensure that the project design considers the unique needs and roles of both women and men. For instance, if the project involves distributing improved livestock breeds or providing training on animal health, the organization will consider how these interventions can benefit women livestock keepers as well as men.
4. *Influencing Resource Allocations*: In the allocation of project resources, such as funds and training opportunities, the organization would prioritize initiatives that address gender disparities. For instance, they might allocate a portion of the project budget to training women in livestock management techniques or ensuring that women have equal access to veterinary services and animal healthcare.
5. *Specialized Inputs*: Throughout the project's lifecycle, the organization would collect and analyse data from a gender perspective. This means tracking the participation and outcomes of both women and men in project activities. For example, they might assess the impact of a livestock vaccination campaign on the livelihoods of women and men separately and adjust as needed. If women face barriers to accessing veterinary services due to cultural norms, the organization might introduce mobile veterinary clinics specifically targeting women livestock keepers.

By implementing these gender mainstreaming strategies in the animal husbandry and livestock sector in India, organizations can work toward a more equitable and inclusive approach that recognizes and addresses the distinct needs and contributions of women and men involved in livestock farming. This can lead to improved livelihoods, increased agricultural productivity, and gender equality in the sector.

3. Suggest a blueprint for gender mainstreaming in livestock development intervention in an area?

Consider a livestock production project in a rural community. The project involves raising goats for both meat and dairy production. Here's how gender mainstreaming could work:

1. Gender Analysis: The project team conducts surveys and interviews to understand the division of labour. They discover that women are responsible for goat care, but men make decisions about which breeds to raise.
2. Equal Opportunities: The project provides training to women on breed selection and management practices, giving those equal opportunities to contribute to decision-making.
3. Participatory Involvement: Women are invited to community meetings where breed selection is discussed, and their input is valued. They express their preferences for specific goat breeds suited to their needs.
4. Gender Budgeting: The project assesses budget allocations and realizes that women are the primary caretakers. It allocates resources for training women in animal healthcare and offers veterinary services closer to their communities.
5. Organizational Training: Livestock extension workers are trained to understand the specific challenges faced by women in goat farming. They offer tailored advice and support.
6. Gender Audits: Periodic audits reveal increased participation of women in decision-making, improved income generation, and better access to resources.

   By implementing these gender mainstreaming initiatives in livestock production, the project aims to create a more equitable and productive environment, where both men and women can benefit equally and contribute to the success of the industry.

4. Explain how technical, political and cultural dimensions of gender mainstreaming in practice can be beneficial in the context of livestock production?

In the context of livestock production, implementing gender mainstreaming involves considering the three dimensions mentioned earlier in the chapter. Let's explore how each of these dimensions can be applied in the context of livestock production, using examples to illustrate their significance:

1. **Technical Dimension:**

   In livestock production, the technical dimension of gender mainstreaming entails ensuring that gender-disaggregated data is collected and used to inform decision-making.

   For instance, let's consider a goat farming project in a rural community. Gender-disaggregated data could reveal that women are primarily responsible for goat care, such as feeding, milking, and veterinary care.

Understanding this data can lead to the development of tailored training programs and resources specifically for women goat farmers, ensuring they have access to the necessary knowledge and tools to improve productivity.

2. **Political Dimension:**

   In the political dimension of gender mainstreaming, it is essential to empower women in livestock production to participate in decision-making processes.

   In our goat farming project, this could mean involving women in community-level discussions about resource allocation, breeding strategies, and marketing. By supporting women's involvement in these decisions, the project not only promotes gender equality but also benefits from the diverse perspectives and expertise that women bring to the table.

3. **Cultural Dimension:**

   The cultural dimension is critical in livestock production, as it can significantly impact traditional gender roles and practices. For example, in a culture where men are traditionally responsible for cattle farming and women for poultry, introducing gender mainstreaming may require sensitively challenging these norms. It might involve community discussions and education programs that encourage both men and women to diversify into various livestock activities. By respecting cultural norms and traditions while promoting gender equality, the project can achieve better outcomes.

   Overall, the successful application of gender mainstreaming in livestock production involves understanding and addressing the technical, political, and cultural dimensions. This approach ensures that gender considerations are integrated into the project from data collection to decision-making, ultimately leading to more equitable and effective livestock production practices

5. Explore how the evolution of gender mainstreaming approach can benefit in the context of the livestock sector?

The evolution of gender mainstreaming approaches in the context of the livestock sector can be understood as under :

a. **Welfare Approach:**

   The welfare approach primarily focuses on providing basic services to women in their roles as caretakers of animals. For example, in rural areas, where women often manage small-scale poultry farming, the welfare approach may be aimed to provide them with training in poultry care, access to veterinary services, and information on nutrition for their poultry. These interventions are designed to improve the health and well-being of both women and the animals they cared for.

b. **Women in Development Approach (WID):**

This approach may recognize that women played a crucial role in livestock management, such as dairy farming. It may aim to integrate women into the sector by providing them with training in modern dairy farming techniques, access to credit and resources, and support for income-generating activities related to dairy products. This approach sought to empower women economically through their active involvement in the dairy value chain.

c. **Gender Efficiency and Gender Empowerment Approaches:**

Gender efficiency analysis may become a tool for considering the roles of both men and women in livestock production. For instance, in large-scale cattle farming, gender efficiency analysis would recognize the distinct roles of men in cattle care and women in dairy processing.

By understanding and addressing these roles, interventions could be better targeted. Additionally, the gender empowerment approach meant working with women as agents of change in the livestock sector. For instance, women in a goat farming project could be involved in decision-making and trained in organizational skills, enabling them to have a more active role in managing their livestock and participating in value addition processes like cheese production.

d. **Gender Mainstreaming Approach:**

In a cattle farming project, gender mainstreaming ensures that women's and men's concerns are equally considered in the design, implementation, and monitoring of the project. This means that not only are women provided access to cattle care training, but they are also involved in decision-making about breeding strategies, cattle health management, and market access. Gender mainstreaming recognizes that women are a target group, and its goal is to promote gender equality across all aspects of livestock production.

By adopting the gender mainstreaming approach in the livestock sector, projects are more likely to address gender disparities effectively, create more inclusive and equitable livestock production systems, and ultimately contribute to the empowerment of women in this field.

SECTION C: SHORT ANALYTICAL QUESTIONS

1. **What are the three key dimensions to consider when applying gender mainstreaming, and why are they important?**

   The three key dimensions are the technical dimension, political dimension, and cultural dimension. They are important because they provide a holistic approach to implementing gender mainstreaming, taking into account data, political advocacy, and cultural context.

2. **How does gender analysis contribute to the effective implementation of gender mainstreaming at the programmatic level?**

   Gender analysis helps identify how and why certain issues affect women and men differently, allowing for the development of targeted solutions and program design that addresses gender disparities.

3. **Why is it essential to allocate adequate resources for gender mainstreaming initiatives within institutions?**

   Adequate resources are necessary to support gender mainstreaming, including conducting gender analysis, developing gender-responsive programs, and providing gender-focused training and capacity-building activities.

4. **What are the potential benefits of gender mainstreaming in terms of social and economic development?**

   Gender mainstreaming can contribute to increased economic growth and improved social indicators by utilizing the skills and talents of all individuals, regardless of their gender.

5. **How can gender mainstreaming help reduce gender-based violence?**

   Gender mainstreaming challenges unequal power dynamics and traditional gender norms, which can contribute to the reduction of gender-based violence.

6. **How does the availability of reliable and sex-disaggregated data contribute to the effectiveness of gender mainstreaming in practice?**

   Reliable and sex-disaggregated data are essential for gender mainstreaming as they provide a factual basis for understanding gender disparities, formulating targeted interventions, and measuring progress. When such data is available, it becomes easier to identify areas where men and women are impacted differently, design programs to address these disparities, and track changes over time to ensure gender equality.

7. **What is the significance of establishing an empowering environment for women to assess their circumstances and voice their preferences in the context of gender mainstreaming?**

   Establishing an empowering environment for women to assess their circumstances and voice their preferences is critical for gender mainstreaming. It ensures that women are not just passive recipients of development initiatives but active participants in defining their needs and priorities. This approach recognizes that women are best positioned to identify the challenges they face and the solutions that will work for them, leading to more effective and sustainable development outcomes.

8. **How do socio-cultural norms and local ideologies influence the execution and interpretation of gender mainstreaming initiatives, and why is it important to consider these factors?**

   Socio-cultural norms and local ideologies play a significant role in shaping the success of gender mainstreaming initiatives. It's important to consider these factors because they can either support or hinder the implementation of gender equality strategies. Understanding local norms and ideologies helps in designing interventions that are culturally sensitive and contextually relevant. Failure to do so can lead to resistance and the ineffectiveness of gender mainstreaming efforts.

9. **What are the key steps involved in gender mainstreaming at the programmatic level, and how do these steps contribute to promoting gender equality?**

   The key steps in gender mainstreaming at the programmatic level include gender analysis, program design, resource allocation, implementation, and monitoring and evaluation. These steps contribute to promoting gender equality by ensuring that programs are tailored to address the specific needs and challenges faced by different genders, that resources are allocated to support these goals, and that progress is measured and adjustments are made as necessary to achieve gender equality outcomes.

10. **How can the establishment of institutional structures, resource allocation, and accountability mechanisms contribute to the successful incorporation of gender considerations into institutions?**

    The establishment of institutional structures, resource allocation, and accountability mechanisms are essential for the successful incorporation of gender considerations into institutions. Institutional structures like gender units provide the necessary framework for gender mainstreaming efforts, while resource allocation ensures that there are adequate funds and personnel dedicated to gender-related initiatives. Accountability mechanisms, such as gender-sensitive indicators, help institutions track and measure progress in achieving gender equality goals, ensuring that these considerations are integrated into the institution's core functions.

11. What are the key drivers for the need for gender mainstreaming in society, and how do they impact various aspects of development and social change?

The key drivers for the need for gender mainstreaming in society include the promotion of gender equality, addressing gender disparities, enhancing policy and program effectiveness, fostering social and economic development, reducing gender-based violence, ensuring human rights, fulfilling international commitments, improving health outcomes, empowering marginalized groups, breaking stereotypes, and achieving long-term social change. These drivers impact various aspects of development and social change by promoting inclusivity, equity, and justice, and by challenging historical and structural inequalities that have affected different genders.

12. What is the primary objective of gender mainstreaming?

The primary objective is to ensure that policies, programs, and projects benefit both women and men, promoting gender equality.

13. What are the central principles of gender mainstreaming?

The central principles include focusing on promoting gender equality, its relevance in all sectors and policy areas, and employing both integrated and targeted actions for achieving gender equality results.

14. How does gender mainstreaming differ from mainstreaming in development policy?

Gender mainstreaming specifically focuses on issues of gender equality, women's rights, and empowerment, while mainstreaming in development policy addresses a wide range of specific development issues like the environment, human rights, disability, and children's rights.

15. What is meant by "the twin-track approach" in gender mainstreaming?

The twin-track approach in gender mainstreaming involves integrating gender considerations into routine processes and procedures while also employing targeted interventions to address specific constraints and challenges faced by women or men.

16. What is the role of gender analysis in gender mainstreaming?

Gender analysis involves collecting and analysing data from a gender perspective to identify and address gender disparities and inequalities in policies, programs, and projects.

17. According to the United Nations Economic and Social Council (ECOSOC), what is the ultimate goal of gender mainstreaming?

The ultimate goal of gender mainstreaming, as per ECOSOC, is to achieve gender equality by making the concerns and experiences of both women and men an integral part of policies and programs.

17. How does gender mainstreaming acknowledge the interconnected roles of both men and women in development?

Gender mainstreaming acknowledges that changing one gender role naturally impacts the other, aiming to reshape unequal interactions and systems generating gender inequality.

18. Why is it essential to consider the six Ws (Who, Which tasks, What activities, When, Where, Why) in livestock extension and advisory services?

Considering the six Ws helps ensure programs are inclusive, culturally sensitive, and tailored to the unique roles and responsibilities of both men and women in the context of livestock management and agriculture.

SECTION D: OBJECTIVE SECTION

Question 1. Multiple Choice Questions

1. Gender mainstreaming involves which of the following dimensions?

a. Technical, financial, and legal

b. Technical, political, and cultural

c. Cultural, social, and economic

d. Political, environmental, and educational

2. What is the primary purpose of gender mainstreaming at the programmatic level?

a. Promoting gender segregation

b. Eliminating gender differences

c. Incorporating gender perspectives into policies and programs

d. Promoting traditional gender roles

3. Which step of gender mainstreaming involves the utilization of insights gained from gender analysis?

a. Gender Analysis

b. Programme Design

c. Resource Allocation

d. Monitoring and Evaluation

4. **In gender mainstreaming, why is it important to allocate sufficient financial and humanresources?**
   a. To promote gender segregation
   b. To support gender-responsive programs
   c. To maintain traditional gender roles
   d. To avoid addressing gender disparities

5. **Which of the following dimensions is NOT mentioned as important in the application of gender mainstreaming?**
   a. Technical Dimension
   b. Economic Dimension
   c. Political Dimension
   d. Cultural Dimension

6. **What is the first step in gender mainstreaming at the programmatic level?**
   a. Monitoring and Evaluation
   b. Programme Design
   c. Resource Allocation
   d. Gender Analysis

7. **Why is it important to allocate resources for gender-focused training programs in gendermainstreaming?**
   a. To create gender disparities
   b. To challenge gender norms
   c. To increase the workload of women
   d. To enhance gender equality

8. **What is one of the key reasons for the need for gender mainstreaming?**
   a. To maintain traditional gender roles
   b. To address gender-based disparities
   c. To exclude marginalized groups
   d. To promote gender-based violence

9. **What is the primary goal of gender mainstreaming?**
   a. Empowering men
   b. Promoting gender equality
   c. Fostering gender disparities
   d. Eliminating women's roles

10. **Gender mainstreaming involves the integration of the gender perspective into which stages of policy processes?**
   a. Only during implementation
   b. Design, implementation, monitoring, and evaluation
   c. Only during evaluation
   d. Design and evaluation

11. In gender mainstreaming, what is the "twin-track" approach?

a. It focuses on integrating gender perspectives in all aspects of policies.

b. It aims to address specific constraints and challenges faced by women and men.

c. It promotes gender disparities.

d. It involves only targeted interventions.

12. What is the role of gender analysis in gender mainstreaming?

a. It perpetuates gender inequalities.

b. It ensures that resources are allocated equally to men and women.

c. It identifies existing inequalities between men and women.

d. It only focuses on one gender's perspective.

13. Which dimension is essential in the application of the Gender Mainstreaming strategy?

a. Economic Dimension

b. Technical Dimension

c. Environmental Dimension

d. Political Dimension

14. How can gender mainstreaming be incorporated into institutions?

a. By excluding gender units and focal points

b. By not allocating resources for gender-focused initiatives

c. By ignoring accountability mechanisms

d. By creating institutional structures, allocating resources, and implementing accountability mechanisms

15. Which is a characteristic of gender mainstreaming?

a. It disregards socio-cultural norms.

b. It primarily relies on specialized women's organizations.

c. It acknowledges the interconnected roles of both men and women.

d. It focuses on creating inequality.

16. What was the primary focus of the Welfare Approach to gender in development?

a. Economic empowerment of women

b. Maternal and child health

c. Women's political participation

d. Access to education for women

17. Which approach aimed to integrate women into the development process by focusing on theirproductive roles?

a. Welfare Approach

b. Women in Development (WID) Approach

c. Gender Efficiency Approach

d. Gender Empowerment Approach

18. Gender mainstreaming seeks to ensure that __________ are integral to the design,implementation, and evaluation of all policies and programs.

a. Women's concerns and experiences

b. Men's concerns and experiences

c. Government officials' interests

d. Economic indicators

Question 2. True or False

1. Gender mainstreaming seeks to incorporate gender perspectives only in policy design.
2. Gender mainstreaming aims to reduce gender-based violence by challenging unequal power dynamics.
3. Gender analysis is a step in the process of gender mainstreaming that involves identifying gender disparities.
4. Adequate allocation of resources is not essential for the effective implementation of gender mainstreaming.
5. Gender mainstreaming primarily benefits men while ignoring the needs of women.
6. Gender mainstreaming requires the availability of reliable data that is sex-disaggregated and derived from a substantial sample size.
7. The political dimension of gender mainstreaming involves advocating for women's elevation as decision-makers and redefining development priorities.
8. Cultural factors have no influence on the execution and interpretation of gender mainstreaming initiatives.
9. Gender mainstreaming involves systematically incorporating gender perspectives into policies, programs, and thematic areas.
10. Gender analysis is the first step in the programmatic level of gender mainstreaming and involves understanding how and why certain issues impact women and men differently.

11. In the program design phase of gender mainstreaming, the insights gained from gender analysis are integrated into program outcomes, indicators, and methods of intervention.
12. Resource allocation in gender mainstreaming ensures that sufficient resources are dedicated to address gender equality considerations throughout the program cycle.
13. Monitoring and evaluation in gender mainstreaming helps establish a factual foundation for strategic decisions related to gender equality and holds institutions accountable for their commitments.
14. The establishment of institutional structures, resource allocation, and accountability mechanisms are key components of incorporating gender mainstreaming into institutions.
15. Gender mainstreaming is a concept aimed at perpetuating traditional gender roles and norms.
16. Gender mainstreaming can contribute to the reduction of gender-based violence by addressing unequal power dynamics and traditional gender norms.
17. Gender mainstreaming aims for short-term social change by addressing gender disparities through targeted interventions.
18. The Welfare Approach assumed that as macroeconomic strategies aimed at modernization and growth benefited men, women would automatically benefit as well.
19. The Women in Development (WID. Approach emerged in response to the growing recognition that women's concerns were neglected in development processes.
20. The Gender Efficiency Approach primarily focuses on what women can do for development at the expense of their own empowerment.
21. Gender mainstreaming was first introduced as a concept after the Fourth World Conference on Women in Beijing in 1995.
22. Gender mainstreaming aims to exacerbate gender inequalities rather than promote gender equality.
23. Gender mainstreaming primarily involves making gender-specific policies for women to address their unique needs.
24. Gender mainstreaming is only relevant at the national policy level and not at the sub-national level.
25. Gender mainstreaming is solely the responsibility of specialized women's organizations.

26. Gender mainstreaming is a fixed and universally applicable blueprint for promoting gender equality.
27. Gender mainstreaming encompasses a technical dimension, a political dimension, and a cultural dimension.
28. Gender mainstreaming is primarily about increasing women's participation as decision-makers.
29. Gender mainstreaming at the programmatic level includes gender analysis, program design, and resource allocation.
30. Gender mainstreaming at the institutional level includes introducing accountability mechanisms and creating institutional structures.

Question 3. Fill in the Blanks

1. Gender mainstreaming seeks to incorporate a ______________ perspective into policies, programs, and projects.
2. The cultural dimension recognizes that socio-cultural norms and ______________ significantly influence the efficacy of gender mainstreaming initiatives.
3. Gender analysis involves examining how and why certain issues impact women and men differently within a specific ______________.
4. In the context of gender mainstreaming, it is crucial to allocate ______________ resources to tackle gender equality considerations.
5. Gender mainstreaming aims to challenge and transform traditional ______________ roles and norms.
6. Gender units and gender focal point systems are institutional mechanisms created to support the integration of gender mainstreaming within ___________.
7. Gender mainstreaming aims to ensure that both ___________ and ___________ concerns and experiences are considered in development initiatives.
8. ___________, a Danish economist, challenged the assumptions of the Welfare Approach in her book 'Women's Role in Economic Development,' highlighting that as men's situations improved, women were increasingly losing status
9. The United Nations declared ___________ the International Year for Women, followed by the International Women's Decade from ___________ to ___________, which increased attention to women's needs and concerns across the development sector.

10. Gender mainstreaming emerged as a response to the frustration with integrating women into development efforts during the _______ and _______.

11. The term "gender mainstreaming" gained recognition and adoption during the Fourth World Conference on Women in _______ in 1995.

12. Gender mainstreaming seeks to account for the differences between women and men when crafting, executing, and assessing _______.

13. Gender mainstreaming involves the integration of a gender equality perspective into every stage and aspect of _______.

14. The primary objective of gender mainstreaming is to ensure that initiatives benefit both women and men, without exacerbating inequalities, but rather promoting _______.

15. Gender mainstreaming is relevant for and should be utilized in _______ and _______.

16. Gender mainstreaming involves both fully integrated and targeted actions for achieving gender equality results, known as the '________' approach.

17. Gender mainstreaming is a globally accepted strategy to promote _______.

18. Gender-related matters should be an integral part of regular institutional activities, not solely relegated to specialized _______ organizations.

19. Gender mainstreaming involves incorporating gender perspectives into policies, programs, and thematic areas at both the _______ and _______ levels.

20. In gender mainstreaming, conducting a _______ is the first step to identify existing inequalities between men and women.

21. Gender mainstreaming involves ensuring equal opportunities for all individuals and implementing gender-specific measures where noticeable _______ persist.

22. Gender mainstreaming strives for equitable participation of both men and women in determining priorities and contributing to the design, development, implementation, direction, and monitoring of _______.

23. Gender budgeting practices are incorporated to evaluate the gender-specific impacts of _______ allocations.

24. Gender mainstreaming involves providing organizational training on tools that facilitate an environment for initiating _______ changes.

ANSWERS

1. Multiple Choice Questions

1	b	Technical, political, and cultural
2	c	Incorporating gender perspectives into policies & programs
3	b	Programme Design
4	b	To support gender-responsive programs
5	b	Economic Dimension
6	d	Gender Analysis
7	d	To enhance gender equality
8	b	To address gender-based disparities
9	b	Promoting gender equality
10	b	Design, implementation, monitoring, and evaluation
11	b	It aims to address specific constraints and challenges faced by women and men.
12	c	It identifies existing inequalities between men and women
13	c	Political Dimension
14	d	By creating institutional structures, allocating resources, and implementing accountability mechanisms
15	c	It acknowledges the interconnected roles of both men and women
16	b	Maternal and child health
17	b	Women in Development (WID) Approach
18	a	Women's concerns and experiences

Answer 2. True and False

S.No.	True or false	S.No.	True or false
1	False	2	True
3	True	4	False
5	False	6	True
7	True	8	False
9	True	10	True
11	True	12	True
13	True	14	True
15	False	16	True
17	False	18	True

19	True	20	True
21	True	22	False
23	False	24	False
25	False	26	False
27	True	28	True
29	True	30	True

Answer 3. Fill in the Blanks

1	Gender
2	Prevalent thoughts and ideologies
3	Setting
4	Financial
5	Gender
6	Institutions
7	Women's, men's
8	Easter Boserup
9	1975,1975-85
10	1970s, 1980s
11	Beijing
12	Policies, programs, and projects
13	Policies, programs, and projects
14	Gender equality
15	All sectors, policy areas
16	Twin-track
17	Gender equality
18	Women's
19	Programmatic, institutional
20	Gender analysis
21	Inequalities
22	Programs
23	Financial
24	Institutional

6

Gender Budgeting in India Tools and Principles

Chapter Overview

This chapter provides an in-depth examination of Gender Budgeting in India, exploring its historical development, principles, and the tools used to promote gender-sensitive budgeting. It delves into the key concepts, institutional frameworks, and the necessity for gender budgeting. The tools for deepening gender budgeting, such as the Five-Step Framework and gender-sensitive checklists, are explained in the context of the Livestock sector. Furthermore, it elucidates the importance of valuing unpaid work and the role of gender budgeting in monitoring policy goals.

SECTION A: THEORY

Gender Budgeting

The inception of the initial gender budgeting effort occurred in 1984 within Australia. This marked a pivotal moment when budgets were recognized as crucial tools in advancing gender parity. Pioneering research into how public budgets influence gender dynamics was conducted. The momentum for gender-based budget analysis surged after the fourth Women's Conference in Beijing in 1995, capturing the interest of governments and non-governmental organizations alike. Following 1995, South Africa took the lead in implementing this approach, followed by similar endeavours in Uganda, Tanzania, Switzerland, and the United Kingdom. Presently, over 100 countries globally have embraced Gender Budgeting. The Government of India integrated Gender Budgeting into its fiscal strategy in 2004-05, and it has since become a routine component of India's annual budget.

Gender Budgeting in India

In India, the perspective of gender-related allocations in public spending has been on the rise since the release of the Committee on the Status of Women's report in 1974.

The Eighth Five-Year Plan (1992-97) marked the initial recognition of the necessity to ensure a dedicated allocation of funds from general developmental sectors to benefit women. The Ninth Five-Year Plan (1997-2002), while reiterating this commitment, incorporated the Women Component Plan as a key strategy. Both the Central and State Governments were directed to allocate "not less than 30 percent of funds/benefits" in all sectors related to women. In 2001, the Government of India adopted the National Policy for the Empowerment of Women. Subsequently, in 2004-05, the Indian government introduced Gender Budgeting as a fiscal approach, which has since become a standard feature of the annual budget. The establishment of Gender Budgeting Cells (GBCs) within all Ministries was mandated in 2005. The introduction of the Gender Budget Statement (GBS) in the Union Budget in 2005-06 marked an annual practice. The Tenth Plan (2002-2007) emphasized the continued dissection of the Government budget to assess its gender-specific impact and emphasized the synergistic role of the Women's Component Plan and Gender Budgeting concepts. The Eleventh Plan (2007-12) aimed to encourage Gender Budgeting and Gender Outcome assessment across all Central and State ministries and departments, with ongoing efforts to establish Gender Budgeting Cells. The Twelfth Plan (2012-17) highlighted "mainstreaming gender through gender budgeting" as a pivotal aspect of addressing Gender Equity among seven key elements. The other six key elements are economic empowerment, social empowerment, health education, violence against women and trafficking, enabling Environment

The plan underscored the intention to strengthen the Gender Budgeting process and expand its implementation to encompass all ministries, departments, and state Governments.

Definitions of Gender Budgeting

Gender budgeting is an application of gender mainstreaming in the budgetary process. It means a gender-based assessment of budgets, incorporating a gender perspective at all levels of the budgetary process and restructuring revenues and expenditures to promote gender equality. In short, gender budgeting is a strategy and a process with the long-term aim of achieving gender equality goals.

(European Institute of Gender Equality)

Gender Budgeting is concerned with gender sensitive formulation of legislation, policies, plans, programmes and schemes; allocation and collection of resources; implementation and execution; monitoring, review, audit and impact assessment of programmes and schemes; and follow-up corrective action to address gender disparities.

(Ministry of Women and Child Development, Govt.of India)

A gender responsive budget is a budget that acknowledges the gender patterns in society and allocates money to implement policies and programmes that will change these patterns in a way that moves towards a more gender equal society. Gender budget initiatives are exercises that aim to move the country in the direction of a gender responsive budget.

(MANAGE, Hyderabad)

Requirements and Principles for Gender Budgeting

Without proper budget allocations the aims and objectives of any plan or program cannot be accomplished. Similarly, if specific budget allocations are not made properly as per the program, the budget will remain unutilised and gender budgeting is no exception. The following are three important principles which need to be adhered for a successful gender budgeting

i. **Political will:** The essential precondition for the success of gender budgeting initiatives is a strong determination from the political sphere. Actively committing to advancing gender equality and acknowledging gender budgeting as a significant strategy in achieving this goal is vital for its effectiveness. Political will involves advocating and raising awareness within relevant political arenas such as national parliaments, regional and local assemblies, consultative bodies, and political parties. If a country has a strong political will to advance gender equality let us say in livestock sector, it means that the government actively supports and endorses gender budgeting in this sector, ensuring that it's discussed during budget meetings and reflected in policies.

ii. **Gender-segregated data**: Gender-segregated data stands as a fundamental requirement for gender budgeting, serving as the essential foundation for evaluating the gender implications of policies at large. National statistics, information systems within ministries, public agencies, and research institutions play a critical role in supplying this foundational data.

    For example, a gender budgeting analysis may reveal that women in rural areas are actively involved in small-scale poultry farming, while men are primarily engaged in large-scale sheep and goat ranching and there are disparities in resource allocation, such as funding and training opportunities, between these two groups. This information can lead to targeted interventions to empower women in poultry farming, such as allocation of funds with access to training and resources to improve their poultry production.

iii. **Emphasizing transparency, partnership and collaboration**: Transparency throughout the budgeting process, as well as broader political decision-making, is a core principle guiding effective gender budgeting. The process of gender budgeting necessitates collaboration between budgetary experts and gender specialists, along with the inclusion of both women and men at every stage. This collaborative approach extends beyond government to encompass civil society organizations, external experts, and various stakeholders.

Collaboration also involves interaction between different government ministries, with particular importance placed on cooperation between the Ministry of Finance and the entities responsible for promoting gender equality. Inclusion of external experts is often beneficial, especially for training government officials and assisting them in conducting gender impact assessments. For instance, a livestock department's budget should be made available to the public, including gender-disaggregated data, so that citizens and civil society organizations can scrutinize and provide feedback. Collaboration between budgetary experts and gender specialists is essential.

Suppose a livestock department is planning a program to improve animal health services, then it is advisable to involve gender experts who can advise on how to make these services more accessible to women who are often responsible for small livestock care. Collaboration also extends to involving men and women from rural communities in the planning and decision-making process. For example, consultations with local women's groups can identify their specific needs for training and resources to improve their participation in the livestock sector. Moreover, cooperation between the Ministry of Finance and the livestock department is crucial. A gender budgeting approach may reveal that the livestock department's budget doesn't adequately address the needs of women livestock farmers. In this case, collaboration between the two entities can lead to a reallocation of funds to ensure gender-responsive programs are implemented. Inclusion of external experts is also beneficial. For instance, an NGO specializing in gender equality can provide training to government livestock officials on conducting gender impact assessments. This ensures that the budgeting process becomes more gender-sensitive and results in improved outcomes for both men and women involved in the livestock sector.

Questions that department/organisation/institution must address during Gender Budgeting

The following are some of the important questions which need to be answered for better clarity and understanding on gender budgeting.

1. What are the Goals and Objectives that the Ministry/Department seeks to achieve? How do they contribute to the larger National Goal of achieving Gender Equality?
2. What are the needs and priorities of women, especially those who are marginalised, in my Ministry/Department's domain of work?
3. Are these presently included and addressed in the Ministry/Department's Policies, Plans, Programmes and Schemes?
4. What activities will the Ministry/Department undertake this year that will reduce gender gaps?
5. What difficulties does the Ministry/Department face in enabling its services to reach women and girls?
6. How can these challenges be addressed?

Questions that department/organisation/institution must address during Gender Budgeting

Example: Animal Husbandry Sector

1. What are the Goals and Objectives that the animal husbandry department seeks to achieve? How do they contribute to the larger National Goal of achieving Gender Equality?

   The animal husbandry department's goal could be to increase the income and livelihood opportunities for rural communities engaged in livestock farming.

   How will this empower rural women who often play a significant role in livestock rearing ?

   Will it enhance their economic independence and social status?

   Will it lead to Gender Equality. If yes how?

2. What are the needs and priorities of women, especially those who are poor, in the animal husbandry department's domain of work?

   Women in rural areas might need better access to livestock healthcare services, training in modern livestock management techniques, and opportunities to participate in decision-making regarding livestock resources.

3. Are these presently included and addressed in the husbandry department's Policies, Plans, Programmes, and Schemes?

   The department may have existing programs to improve livestock healthcare, but it's important to assess whether these programs adequately address the specific needs and priorities of rural women.

4. What activities will the husbandry department undertake this year that will reduce gender gaps?

   The department could launch a program in assessing the specific skills required by women to improve their capacity in livestock production and management and providing appropriate training and support to them, ensuring their access to inputs needed to employ the skills acquired and encouraging their participation in livestock-related decision-making aspects.

5. What difficulties do the husbandry departments face in enabling its services to reach women and girls?

   Challenges may include limited awareness among women about available services, lack of resources allocated specifically for women in livestock, and social or cultural barriers that restrict women's participation in addition to lack of women extension advisors.

6. How can these challenges be addressed?

   To address these challenges, the department can run awareness campaigns targeting women, allocate a portion of the budget to support women in livestock, and engage with local communities and leaders to change traditional gender norms that restrict women's involvement in the sector.

   These questions help the animal husbandry department/ministry consider the gender dimension in its budgeting and planning, ensuring that their policies and programs promote gender equality and empower women in the sector

Tools for deepening gender budgeting

UN-HABITAT on the "Gender Budgeting Handbook for Government of India and Departments" provides valuable insights into several essential tools crucial for the successful implementation of gender budgeting. Some of these tools are discussed below

1. The Five-Step Framework for Gender Budgeting

Debbie Budlender and Rhonda Sharp originally gave a five step Framework for Gender budgeting as discussed below:

Step 1: An analysis of the situation for women and men and girls and boys (and the different sub-groups) in each sector.

A thorough analysis is conducted to understand the existing gender disparities and dynamics within the Livestock sector. Data may reveal that women in rural areas are actively involved in livestock rearing, but they face barriers such as limited access to veterinary services, market opportunities, and training compared to men. This step helps identify the gender-related issues in the Livestock sector. It may also reveal some areas where the extent of involvement of men and women is almost same. For example, both men and women are involved in milking of cows.

Step 2: An assessment of the extent to which the sector's policy addresses the gender issues and gaps described in the first step.

Once the gender issues are identified, the next step is to evaluate the extent to which the sector's policies address these issues. For example, it may be found that existing Livestock sector policies do not explicitly consider the needs and challenges faced by women in accessing veterinary services or in livestock credit services. The policy may be to invite men only for on campus training program to improve their knowledge and skills in performing certain activities which are primarily in the domain of women.

Step 3: An assessment of the adequacy of budget allocations to implement the gender-sensitive policies and programmes identified in step 2.

Application in animal husbandry: After identifying gender-sensitive policies, the budget allocations are assessed to ensure that they are adequate for implementing these policies effectively. In Livestock sector example, it might be revealed that there is a significant disparity in the budget allocation for women's training and support compared to other livestock sector activities.

Step 4: Monitoring whether the budget ear marked was spent as planned, what was delivered and to whom.

Application in animal husbandry: This step involves monitoring whether the allocated funds were spent as planned and tracking what was delivered and to whom. In Livestock sector example, it could involve monitoring whether the budget allocated for women's training programs was utilized as intended and whether the training reached the targeted women in rural areas. It is possible that the finance allocated for training women on preparation of balance feed may be diverted to some other activity.

Step 5: An assessment of the impact of the policy/ programme/scheme and the extent to which the situation described in step 1 has changed.

Application in animal husbandry: The final step is assessing the impact of the Livestock sector policies and budget allocations on gender disparities. For instance, after implementing gender-sensitive policies and adequately allocating the budget to address women's needs, an impact assessment may reveal an increase in women's effective participation in livestock activities with improved capacity, better access to veterinary services and ultimately improved family income.

In this way, the Five-Step Framework for Gender Budgeting ensures that gender issues and disparities within the Livestock sector are systematically addressed through policy formulation, budget allocation, implementation, and monitoring, ultimately leading to positive changes and greater gender equality in the sector.

2. Participatory planning and budgeting (can happen at all five steps)

Participation of women (and men) is important while planning and budgeting because women's priorities in the use of public funds may be different from those of men. It is not enough to say that women's needs and concerns have been considered. Women must be treated as equal partners in decision making and implementation rather than only as beneficiaries.

3. Spatial Mapping (Corresponds to step 1 of Five stage framework of Gender Budgeting)

Spatial mapping in the context of gender budgeting involves visualizing and analysing geographic or spatial data to understand how budget allocations and the implementation of programs and schemes affect different regions or areas, particularly in terms of gender disparities. It helps identify areas with varying gender-related challenges, opportunities, and needs. Spatial mapping can provide insights into resource allocation, access to services, and the impact of policies on different groups of people in different geographic locations.

Spatial mapping in the livestock sector, as in other sectors, helps governments and policymakers make informed decisions by identifying geographic areas where gender disparities are most pronounced and where targeted interventions are needed to promote gender equality in budget allocations and resource distribution.

4. Gender Appraisal for New Programme and Schemes (Corresponds to step 2 of Five stage framework of Gender Budgeting)

All new programs, projects, and schemes requesting funds should undergo a gender appraisal. This involves evaluating them through a gender lens to identify potential impacts on different genders.

For example, the government is launching a new program in the animal husbandry sector aimed at improving livestock health services. A gender appraisal would involve assessing how this program may impact men and women differently. For instance, women are often responsible for small livestock care. The appraisal would consider whether the program's design is inclusive and accessible to women, addressing their specific needs in livestock management.

5. Guidelines for Gender-Sensitive Review (Corresponds to step 2 of Five stage framework of Gender Budgeting)

The Ministry of Women and Child Development has developed specific guidelines, Checklists I and II. Checklist I focus on beneficiary-oriented programs targeting women, while Checklist II covers mainstream sectors. These guidelines aid in reviewing public expenditure and policies from a gender perspective, identifying constraints, and suggesting corrective actions.

For example, using the Ministry of Women and Child Development's guidelines, the government reviews the livestock health program through Checklists I and II. Checklist I ensure that the program consciously targets and benefits women involved in livestock care. Checklist II, on the other hand, helps assess the gender impact of the program on mainstream sectors, addressing any disparities in resource allocation or service distribution.

6. Gender-Based Profile of Public Expenditure (Corresponds to step 3 of Five stage framework of Gender Budgeting)

Preparation of a gender-based profile facilitates the review of schemes and public expenditure. It highlights the gender component of expenditure, physical targets, and the trend of targeted expenditure in terms of male/female beneficiaries. This exercise reveals the extent to which budgeting is gender-responsive and may identify constraints like the unavailability of gender-relevant data. The government prepares a gender-based profile for the animal husbandry sector, analyzing the budget allocation and physical targets in terms of gender. This reveals whether the funds allocated to livestock health services adequately consider the needs of both men and women, and if there's equitable access to the program's benefits.

7. Outcome Budget (Corresponds to steps 4 and 5 of Five stage framework of Gender Budgeting)

The Charter for Gender Budget Cells suggests incorporating the results of gender analysis of Ministry budgets into their outcome/performance budgets. This involves aligning budget allocations with intended outcomes and assessing how gender considerations contribute to overall program success. Gender Budget Cells within the Ministry of Animal Husbandry incorporate gender analysis into the outcome budget. This means assessing how the program's outcomes contribute to gender-related objectives, such as improving the economic empowerment of women involved in livestock farming.

8. Impact Analysis (Corresponds to steps 5 of Five stage framework of Gender Budgeting)

Beyond monitoring the reach of funds, impact analysis evaluates the effectiveness of government programs. Various methods such as impact assessments, evaluations, and field-level surveys are employed. Results of impact assessments from a gender perspective help identify barriers to women's access to public services and expenditure. They may also reveal cases where although women were reached, desired changes in their situations did not occur, prompting the need for adjustments in program design and implementation modalities.

After implementing the livestock health program, an impact analysis is conducted. This involves assessing whether the program has positively impacted both male and female livestock farmers. For example, the analysis may reveal that while women were reached by the program, there might be challenges in ensuring sustained improvements in their livelihoods. Adjustments in the program's design or implementation can then be made based on these findings.

Checklist I for Gender-Specific Expenditure in Animal Husbandry

Planning and Budgeting:

- List schemes and programs in animal husbandry that are gender-specific, where the targeted beneficiaries are primarily women.
- Briefly indicate activities undertaken under the program for women. In the animal husbandry sector, this may involve training programs, access to veterinary services, or financial support for women engaged in livestock farming.
- Indicate expected output indicators, such as the number of women beneficiaries, increase in women's employment in animal husbandry, and improvements in resources/income/skills of women post-program implementation.
- Quantify allocation of resources in the annual budget for gender-specific animal husbandry programs and specify physical targets.
- Assess the adequacy of resource allocation by considering the population of targeted beneficiaries, the historical trend of past expenditure, and the specific needs of women in livestock farming.

Performance Audit:

- Review actual performance, both physically and financially, against annual targets. Identify constraints in achieving targets, such as the need for strengthening delivery infrastructure or capacity building for women in the animal husbandry sector.

- Conduct a reality check by evaluating program intervention and assessing the impact on women engaged in animal husbandry. This may include comparing the status of women before and after program implementation.
- Compile a trend analysis of expenditure and output indicators, along with impact indicators, to gauge the effectiveness of gender-specific animal husbandry programs.

Future Planning and Corrective Action:

- Address constraints identified in the performance audit, considering the specific needs and challenges faced by women in the animal husbandry sector.
- Establish the requirement of resources based on the population of targeted women beneficiaries and the magnitude of perceived problems in animal husbandry.
- Review the adequacy of available resources, both financial and physical, including trained personnel in animal husbandry.
- the review, ensuring that gender-specific considerations are integrated into future planning for animal husbandry.

Checklist II for Mainstream Sectors in Animal Husbandry

- List all animal husbandry programs entailing public expenditure, with a brief description of activities involved.
- Identify the target group of beneficiaries/users in animal husbandry, categorizing them by gender (male/female) where feasible.
- Evaluate the possibility of undertaking special measures in animal husbandry to facilitate access to services for women, such as through affirmative action, expanding women-specific services, or adopting gender-friendly practices.
- Analyze the employment pattern in rendering animal husbandry services from a gender perspective, exploring avenues to enhance women's recruitment.
- Focus on special initiatives to promote the participation of women in the animal husbandry sector, either in employment or as users of services.
- Indicate the extent to which women are engaged in decision-making processes within the animal husbandry sector and organizations, and initiate actions to correct gender biases and imbalances.

Source: Gender Budgeting Handbook for Government of India and Departments by UNHABITAT for diverse gender perspectives and avoid reinforcing traditional gender roles or biases.

Checklist for integrating gender/gender budgeting into new programs, projects, and schemes (PPS) to the context of the animal husbandry sector

I. Background and Justification:

Following five checklists would help to understand whether gender /gender budgeting has been incorporated.

1. **Inclusion of Gender in Context Analysis:** Check whether gender is a part of the context analysis in the Animal Husbandry PPS. Understanding the roles, contributions, and dynamics of both men and women in animal husbandry is crucial for effective program implementation.
2. **Arguments for Gender Mainstreaming and Equality:** The PPS should include arguments advocating for gender mainstreaming and equality within the animal husbandry sector. This involves acknowledging the importance of both genders, addressing disparities, and promoting equal opportunities and access to resources.
3. **Sex-Disaggregated Data in Background Information:** It's essential for the background data to be sex-disaggregated, distinguishing between male and female stakeholders. This helps in understanding the specific needs, challenges, and contributions of men and women involved in animal husbandry.
4. **Identification of Different Needs and Concerns:** The PPS should identify and address the diverse needs and concerns of men and women, as well as girls and boys, within the animal husbandry sector. This promotes inclusivity and ensures that interventions are tailored to meet the specific requirements of various gender groups.
5. **Gender-Sensitive Language and Avoidance of Stereotypes:** Language used in the PPS should be gender-sensitive and free from stereotypes. It should reflect respect

II. Goal/Objective:

a. Does the goal or objective reflect the needs of women and men in animal husbandry?
b. Does it aim to meet Practical Gender Needs (PGN) or Strategic Gender Interests (SGI)?

Assess whether the stated objectives of the animal husbandry program align with the needs and contributions of both men and women in the sector. Determine whether the objectives of the animal husbandry program address immediate practical needs or also aim to bring about strategic changes that promote gender equality and women's rights.

III. Target Group/Stakeholders

- Are women and men both going to benefit from the Animal Husbandry PPS?
- Is there a need for affirmative action to ensure women benefit?

  Evaluate if the benefits of the program are equitable and accessible to both men and women engaged in animal husbandry. Consider whether affirmative action, such as quotas or reservations, is necessary to ensure women's active participation and benefits in the animal husbandry program.

IV. Strategy and Activities

- Are the strategies and activities of the Animal Husbandry PPS gender-sensitive?
- What are the constraints to women benefiting? Does the PPS address these constraints?
- Will the PPS entail an additional burden on women? If so, what steps will the PPS take for men to share the burden of women's traditional roles?
- If technology is involved, is it women-friendly and appropriate for women?

Examine whether the planned strategies and activities in the program account for gender differences and are inclusive of both male and female participants in animal husbandry. Identify and address any obstacles or challenges that women may face in benefiting from the animal husbandry program, ensuring that the PPS actively addresses and reduces these constraints. Consider whether the program introduces any additional responsibilities for women and, if so, how it plans to involve men in sharing these responsibilities to avoid reinforcing traditional gender roles in animal husbandry. Evaluate whether any technological components of the animal husbandry program are women-friendly and suitable for the context in which women operate in the sector.

V. Budgeting for Equality

- Has sufficient budget been allotted for each of the components of the Animal Husbandry PPS?
- Has the PPS budgeted for gender training?
- Is the budget sufficiently disaggregated to ensure that gender concerns are adequately addressed?
- Has the PPS budgeted for monitoring?

Verify if the budget allocations for the animal husbandry program adequately cover each component, ensuring that gender considerations are incorporated. Confirm whether the budget includes provisions for gender training, acknowledging the importance of capacity building to address gender-related issues in animal husbandry. Ensure that the budget breakdown provides clear details on how funds are allocated to address specific gender concerns in the animal husbandry program. Check if the budget includes provisions for monitoring activities related to gender and if sufficient resources are allocated to ensure effective oversight of gender-related aspects in animal husbandry.

VI. Indicators for Measuring Outcomes and Outputs

- What are the indicators for measuring progress on outcomes and outputs? Are they sex-disaggregated and gender-sensitive?
- Are the indicators SMART - specific, measurable, accurate, relevant, and time-bound?
- Do the indicators measure progress in achieving strategic gender interests (SGIs) as well as practical gender needs (PGNs)?

Confirm that the indicators meet the SMART criteria, ensuring that they provide clear and targeted insights into the gender-specific outcomes and outputs of the animal husbandry program. Evaluate whether the indicators used to measure progress in the animal husbandry program are sensitive to gender differences and are disaggregated by sex. Check if the indicators consider both strategic gender interests and practical gender needs in animal husbandry, addressing immediate requirements while also contributing to long-term gender equality goals.

VII. Monitoring

- Has the PPS built in participatory on-going monitoring, involving women? What is the frequency? And are the monitoring tools (formats, visit timings, etc.) women-friendly?
- Examine whether the monitoring strategy for the animal husbandry program includes the active involvement of women, considering the frequency and women-friendly aspects of monitoring tools.
- Does the monitoring strategy look at both content and process?

Confirm that the monitoring strategy evaluates both the content and the process of the animal husbandry program, recognizing the importance of assessing not only outcomes but also the effectiveness of implementation processes.

VIII. Evaluation

- Has the PPS provision for a mid-term (after 2 or 3 years) and an end-term (if the PPS is for a fixed duration) evaluation?
- Does the evaluation design allow for
  (a) the differential impact of the PPS on men and women to come out clearly;
  (b) women to be part of the evaluation team; and
  (c) perspectives and feedback from women beneficiaries to be obtained first-hand and not through male family members?

Ensure that the animal husbandry program includes provisions for both mid-term and end-term evaluations, allowing for a comprehensive assessment of its impact over time. Confirm that the evaluation design for the animal husbandry program considers the differential impact on men and women, includes women in the evaluation team, and gathers direct feedback from women beneficiaries to ensure a nuanced and gender-sensitive assessment.

Source: Gender Budgeting Handbook for Government of India & Departments by UNHABITAT

Institutional Framework for Gender Budgeting in India

Nodal Department for Gender Budgeting

At central level, Ministry of Women and Child Development is responsible for promoting gender budgeting. At state level women and child development, social welfare, finance or planning department can be the nodal department. These departments aim to build capacities of officials involved in the policies, programmes and projects to undertake gender sensitive planning, budgeting and implementation.

Different Stakeholders in Gender Budgeting

i. Ministry of Finance (both at the centre & state)
ii. Ministry of Women and Child Development/Social welfare department
iii. Women Self Help Groups
iv. Women co-operatives and women farmer producer organisations
v. Comptroller and Audit General of India/Local Audit Departments
vi. Sectoral Ministries like Minister of Agriculture / Health / Education/ labour
vii. Researchers, Economists and Statisticians
viii. Civil society organisations and budget groups

ix. Parliamentarians, budget committee of both houses and other representatives of the people at district and sub-district levels

x. Media

xi Development / Donors

Scope of Gender Budgeting

Gender Budgeting can be done for:

- The whole budget; or
- Impact of expenditure of selected departments or programmes; or
- Gender-sensitive design of new programmes and projects; or
- Assessment of selected forms of revenue and changes in tax system; or
- New legislation

Gender Budget Cell

It is an institutional mechanism to facilitate the integration of gender analysis into the government budget, to tackle gender imbalances and promote gender equality. At government of India level,57 ministries and departments have already setup gender budgeting cells to pursue gender budgeting

Gender budget Statement

A Gender Budget Statement (popularly known as Statement 20) was introduced in Union Budget 2005-06. It is a reporting mechanism that can be used by Ministries/Departments to review their programmes from a gender lens and is an important tool for presenting information on the allocations for women. The Gender budgetary allocations in India have two parts: Part A includes schemes with 100% allocation for women such as the widow pension scheme, girls' hostel scheme and maternity benefit scheme; and Part B with schemes allocating 30% to 99% of funds for women, such as the mid-day meals programme, the rural livelihoods mission and the biogas programme. Since its inception, the gender budget has been dominated by allocations under Part B, accounting for at least two-thirds of the total Gender Budget.

Why does Gender Budgeting Focus on Women?

Around the world, Gender Budgeting tends to focus on women because:

i. Women make up nearly two thirds of the global illiterate population.

ii. Within developing nations, maternal mortality remains a significant cause of death for women in their reproductive years.

iii. Women are notably underrepresented in decision-making roles, both in governmental and corporate spheres, particularly at senior levels.

iv. Women's economic contributions differ greatly from men's, often involving informal and lower-status employment, resulting in unequal pay for equivalent work.
v. Women are primarily responsible for unpaid household tasks such as childbirth, upbringing, and caregiving for dependents and other members of society.

Need for Gender Budgeting

The basis for implementing gender budgeting emerges from acknowledging that national budgets have distinct effects on males and females because of how resources are distributed. Although women make up 48% of India's population, they trail men across several societal indicators such as well-being, education, economic prospects, and more. For example, India ranked 70th out of 77 countries on the Female Entrepreneurship Index and 57th out of 65 nations on the Master Card Index on Women Entrepreneurship. Women entrepreneurship accounts for 20% of the Micro, Small and Medium Enterprises (MSME) sector in India. This substantiates the need for focused consideration due to their susceptibility and constrained resource availability. The approach by which government budgets allot resources holds the capability to reshape these gender disparities. Considering this, Gender Budgeting has been advocated to foster gender mainstreaming, recognizing its potential for transforming these imbalances.

1. Achievement of Gender Equity/Equality

The Constitution of India not only confers parity for women but also authorizes the government to implement affirmative actions in support of women. Nonetheless, significant disparities persist between the ideals outlined in the Constitution, laws, regulations, blueprints, initiatives, and related mechanisms on one side, and the actual circumstances faced by women and girls on the other. Gender budgeting serves as a mechanism to ensure that the government's commitment towards gender parity and the progression of women to receive the necessary financial backing. Budgets and policies can wield varying impacts on men and women, potentially exacerbating or mitigating gender disparities, sometimes even when unintended. Women, men, girls, and boys possess differing priorities and structure their lives dissimilarly, leading to uneven access to both societal benefits and economic provisions.

2. Monitoring the achievement of policy goals

Gender Budgeting is a tool to monitor the achievement of the goals of the National Policy for Empowerment of Women 2001 and other policy goals in a gender-aware manner

3. Valuing Unpaid Work

The traditional notion of an economy does not account for unpaid tasks such as childcare, household chores (including cooking, cleaning, fetching water), elderly care, and voluntary contributions to civil society. It's important to recognize that the combined effort of unpaid labor alongside monetary economic activities constitutes the overall economic productivity of a society. As a result, strategies should be developed to provide support for individuals, both women and men, who actively engage in unpaid work, thereby reducing their burden. As an illustration, within the context of the state-level National Rural Drinking Water Programme (NRDWP), nearly half (47%) of the funds are designated for coverage. This allocation is crucial for alleviating the burden, particularly on women and girls, associated with tasks like fetching drinking water from far away locations. Additionally, it addresses issues like malnutrition and enhances the time available for education and leisure. Moreover, this approach helps prevent potential contamination linked to obtaining water from distant sources.

Need for Gender Budgeting in Animal Husbandry

Gender budgeting is an essential approach to public financial management that seeks to integrate gender perspectives into the entire budgetary process. It ensures that government expenditures and policies are designed in a way that addresses gender disparities and promotes gender equality. Gender budgeting is not limited to issues directly related to human beings; it can also play a crucial role in sectors like animal husbandry. In the context of animal husbandry, gender budgeting becomes relevant since gender roles and responsibilities often influence the way resources are allocated, accessed, and utilized in this sector. Here's an example to illustrate the need for gender budgeting in animal husbandry:

Imagine a rural community where dairy farming is a significant source of income and nutrition. In this community, both men and women are involved in various aspects of dairy farming, such as milking, feeding, and selling dairy products. However, due to traditional gender norms, there might be disparities in the access to and control over resources.

i. **Access to Resources**: In many societies, men might have easier access to resources such as land, credit, and technology. This could lead to unequal opportunities in dairy farming. Gender budgeting would involve assessing these disparities and ensuring that both men and women have equal access to resources required for dairy farming. Funds could be allocated to provide women with better access to land, training, and modern farming equipment.

ii. **Training and Capacity Building**: Gender roles might lead to differences in the types of training and capacity building that men and women receive. For instance, women might be primarily responsible for milking cows but might not receive training in clean milk production practices. Gender budgeting would involve allocating funds for training programs that are tailored to the needs of both men and women, ensuring that they can both contribute effectively to the dairy farming enterprise.

iii. **Marketing and Value Addition**: Women might be more involved in processing dairy products and selling them locally, while men might be involved in larger-scale commercial dairy operations. Gender budgeting would support initiatives that help women add value to their dairy products and access broader markets, thereby increasing their income-generating potential.

iv. **Time and Labour Distribution**: Women often bear the burden of unpaid care work, which includes activities like fetching fodder, feeding animals, making dung cakes and cleaning. Gender budgeting would consider the time and effort spent by women on these tasks and allocate funds for mechanization or labour-saving technologies that can reduce their workload or drudgery.

By implementing gender budgeting in animal husbandry, policymakers and stakeholders can address these disparities and ensure that both men and women have equal opportunities and benefits in the sector. This not only promotes gender equality but also enhances the overall efficiency and productivity of the animal husbandry sector.

Chapter Summary

- Gender budgeting is a strategic tool aimed at advancing gender equality by assessing how budget allocations impact men and women differently.
- Gender budgeting gained prominence globally in the late 20th century, with India incorporating it into its annual budget in 2004-05.
- The key principles of gender budgeting include political will, gender-segregated data, transparency, partnership, and collaboration.
- Tools for deepening gender budgeting include the Five-Step Framework, participatory planning and budgeting, spatial mapping, and gender-sensitive checklists.
- The Five-Step Framework includes analysing gender disparities, assessing policy responses, budget allocation, monitoring and implementation, and impact assessment.

- Gender-sensitive indicators are used to measure changes in women's participation, ownership, decision-making influence, and the transformation of gender norms in sectors like animal husbandry.
- Gender budgeting is critical to address gender disparities, promote equity, and measure policy effectiveness in various sectors.

  It helps monitor policy goals, value unpaid work, and ensure that national budgets promote gender equality, which aligns with the principles of the Indian Constitution and international gender equity standards.

SECTION B: DESCRIPTIVE QUESTIONS

1. Discuss the five-Step framework for gender budgeting citing an example in animal husbandry?

The five-Step framework for gender budgeting in animal husbandry is explained as under

Step 1: Analysis of the Situation

In the animal husbandry sector, it's important to analyse the roles and responsibilities of women, men, girls, and boys within the sector. Traditionally, certain tasks might be considered more suitable for one gender, such as women being responsible for small-scale poultry farming while men focus on larger livestock. This step involves examining the division of labour, access to resources, decision-making power, and other gender-related dynamics within the sector.

Step 2: Assessment of Sector Policy

Review the existing policies and regulations in the animal husbandry sector.

Are they gender-sensitive?

Do they address the inequalities and gaps identified in step 1?

For instance, a gender-sensitive policy might ensure that women have equal access to training, resources, and markets. It could also encourage diversification of livestock production to help women engage in more profitable activities.

Step 3: Adequacy of Budget Allocations

Examine the budget allocated for implementing the gender-sensitive policies identified in step 2.

Is there enough funding to support initiatives that promote gender equality?

For instance, funding might be needed to provide off campus training to women in livestock management techniques or to set up women's cooperatives for collective marketing.

Step 4: Monitoring and Implementation

This step involves closely monitoring whether the allocated budget was used as planned.

Were the training programs conducted?

Were resources distributed equitably?

Monitoring also involves tracking who benefited from the programs –Whether women, men, girls, and boys all able to access the opportunities as intended?

Step 5: Impact Assessment

After implementing the gender-sensitive policies and programs, assess the impact on the animal husbandry sector and gender equality.

Have women been able to take on more diverse roles?

Have income levels improved for both women and men?

Has decision-making become more inclusive?

Has it reduced drudgery forwomen?

This step helps determine whether the changes have led to a more equitable situation compared to the initial analysis in step 1.

Overall, this process ensures that gender issues are considered at every stage of policy formulation and implementation in the animal husbandry sector, leading to more inclusive and effective outcomes.

2. Explain Spatial mapping, a tool in Gender budgeting with example in Livestock sector?

The government allocates budget for livestock-related programs and initiatives, such as animal health services, livestock extension programs, and livestock market development. Spatial mapping can help analyse the impact of these allocations in various regions and reveal gender disparities. Here's how it works:

i. Data Collection: Gather data on the distribution of livestock resources, such as the number and type of livestock (e.g., cattle, goats, poultry), in different geographic regions of the country.

ii. Gender Analysis: Conduct a gender analysis to understand how men and women are involved in the livestock sector. This includes looking at gender roles, access to resources, and participation in livestock-related activities.

iii. Resource Allocation: Map the budget allocation for the livestock sector to specific regions. This will show how much funding is directed to different areas.

iv. Spatial Mapping: Create maps that visually represent the distribution of livestock resources, budget allocations, and gender roles in different regions. For example, the map might show that certain regions have a higher number of women involved in poultry farming, while others have more men engaged in cattle rearing.

v. Identifying Disparities: Analyse the spatial maps to identify disparities. You might find that regions with a higher concentration of female poultry farmers receive less budget allocation compared to regions dominated by male cattle farmers.

vi. Trouble Spots and Special Attention: Spatial mapping can highlight "trouble spots" where gender disparities are most significant. In the livestock sector, this might mean identifying regions where women face significant challenges in accessing livestock-related resources and services, or where budget allocations are insufficient to address these disparities.

vii. Policy Recommendations: Based on the spatial mapping results, policy recommendations can be made to address these disparities. For example, the government might allocate additional resources to regions with high concentrations of female poultry farmers or implement gender-sensitive training programs to empower women in male-dominated livestock activities.

In summary, spatial mapping in the livestock sector, as in other sectors, helps governments and policymakers make informed decisions by identifying geographic areas where gender disparities are most pronounced and where targeted interventions are needed to promote gender equality in budget allocations and resource distribution.

3. Provide examples of successful gender budgeting initiatives in different countries?

There are several examples of successful gender budgeting initiatives in different countries. For instance, in India, the Disha scholarship scheme of the Department of Science and Technology provides support for women scientists, helping them to re-enter the workforce after a break in their career paths due to social responsibilities. In South Africa, the government has implemented a gender-responsive budgeting process that has led to increased funding for programs that address gender-based violence and support women's economic empowerment. In Australia, the government has established a Women's Budget Statement that provides a gender analysis of the budget and outlines measures to promote gender equality. These are just a few examples of successful gender budgeting initiatives that have been implemented in different countries.

4.Discuss gender Budgeting and gender auditing with example in animal husbandry ?

Gender auditing constitutes a component of the gender budgeting procedure. Gender auditing takes place after the budget's execution. It involves scrutinizing financial allocations, observing trends over a period, evaluating percentage distributions, and comprehensively analysing the established systems, employed processes, as well as the results and effects of budget allocations, all from a gender perspective. Gender auditing functions as a compliance measure for gender budgeting. Let's understand gender budgeting & auditing in animal husbandry. Gender budgeting includes:

1. **Allocation Analysis:** Examining how much of the budget is allocated to programs that directly benefit women in animal husbandry, such as distribution of goats to women, training programs exclusively for women livestock farmers or providing them access to veterinary services.
2. **Gender-Sensitive Indicators:** Developing indicators to measure the impact of the budget on gender-related outcomes. For example, tracking the increase in number of all women cooperatives, women self-help groups, women entrepreneurs in livestock enterprises etc.
3. **Targeted Initiatives:** Identifying areas where gender disparities exist. If it's found that women are less likely to participate in on campus training programmes to enhance women's capacity in goat management, the budget might allocate funds specifically to organise off campus training programmes for the convenience of women.
4. **Inclusion of Marginalized Groups:** Ensuring that marginalized groups within the context, such as female-headed households in rural areas, are considered in budget allocations to provide them equal opportunities in animal husbandry.

After the budget has been executed, gender auditing comes into play:

5. **Financial Review:** The Auditors would analyse how the allocated funds were spent in terms of gender-specific programs and initiatives. Whether the funds were fuly utilised the purpose for which the funds were allocated or diverted to some other activities?
6. **Outcome Evaluation:** The auditors would assess the outcomes of the budget execution in terms of gender-related goals. Have the targeted indicators shown improvements? Have gender disparities reduced?
7. **Process Analysis:** The auditors would evaluate whether the processes for implementing gender-specific programs were effective and efficient. Did women face any barriers in accessing the benefits? Were the programs designed to be inclusive?

8. **Feedback from Beneficiaries:** Gathering feedback from women and men who were beneficiaries of the programs. Did the budget allocations address their needs? Were there any unforeseen challenges?
9. **By conducting gender auditing:** the government department can ensure that the budget allocations have indeed benefited women and men equitably within the animal husbandry sector. It is a way to hold the department accountable for its gender-focused goals and to adjust for future budget cycles based on lessons learned from the audit

5. Discuss the distinction between Outputs and Outcomes with example in animal husbandry?

Outputs denote the quantifiable measure of goods and services generated through an activity within a scheme or program. Outcomes, on the other hand, represent the results produced by various governmental initiatives and interventions. To illustrate, establishing dairy cooperatives signifies an output, while the rise in sale of milk and income through sale of milk constitutes an outcome.

Outputs in Animal Husbandry: An animal husbandry program might aim to increase the milk production of dairy cows. In this case, the outputs would be the measurable quantities of milk produced by each cow because of the program. For example, if the program involves providing improved feed and healthcare to the cows, the increased milk yield per cow would be an output.

Outcomes in Animal Husbandry: Taking the same example of the animal husbandry program focusing on increasing milk production, the outcomes would represent the broader impacts of the program on the community and economy. An outcome could be an increase in the income of local dairy farmers due to the higher milk yields. This could lead to improved livelihoods and economic growth in the region. Additionally, an outcome might be an improvement in the overall nutritional status of the community, as the increased milk production provides more dairy products for consumption.

The distinction between outputs and outcomes is important for evaluating the effectiveness and impact of programs. While outputs provide a measure of the direct results of activities, outcomes focus on the broader changes and benefits that result from those activities.

6. Discuss Gender Neutral and Gender Sensitive Sectors

While no sector is entirely gender-neutral, certain sectors like defense, power, trade, transport, and commerce are perceived as such due to their wide-ranging applicability, and their revenue and expenses cannot be divided based on gender. Traditionally, gender-sensitive sectors are those in which the beneficiaries, primarily women, are clearly identifiable. Examples include health, education, labor, and rural development.

7. Explain Gender-Sensitive Indicators with example?

Gender-sensitive indicators aim to measure relative changes for both women and men. These indicators contribute to promoting gender equality or gauge shifts in gender-related equity matters. Gender-sensitive indicators necessitate collecting data disaggregated by sex, covering both quantitative and qualitative aspects. Quantitative indicators involve numerical measurements of change, such as increases in women's or men's literacy levels, income levels, and participation in training. Qualitative indicators endeavor to gauge attitudes, status, and empowerment, focusing on capturing processes and qualitative distinctions. The examples below explain them.

Improving Women's Participation in Poultry Farming

Quantitative Indicator: Increase in Women's Ownership of Poultry Farms

Percentage change in the number of women-owned poultry farms over a specific time (before and after implementation of a programme).

In many societies, women often have limited ownership and control over productive resources, including land and livestock. Promoting women's ownership of poultry farms can contribute to their economic empowerment and challenge traditional gender roles.

Qualitative Indicator: Empowerment of Women in Poultry Decision-makingQualitative assessment of women's involvement in key decision-making processes related to poultry farming, such as investment decisions, flock management, and marketing strategies.

Gender-sensitive indicators should focus on quantitative changes and on the processes that lead to empowerment. Assessing women's active participation in decision-making within the poultry sector reflects their changing roles and influence in traditionally male-dominated activities.

By utilizing both quantitative and qualitative indicators, this example illustrates how gender-sensitive indicators can be applied to animal husbandry. These indicators help track changes in women's participation, ownership, decision-making influence, and the transformation of gender norms within the context of poultry farming.

SECTION C: SHORT ANALYTICAL QUESTIONS

1. **What is gender budgeting, and what is its overarching goal?**

   Gender budgeting is the incorporation of a gender perspective in the budgetary process to promote gender equality. Its overarching goal is achieving gender equality.

2. **What are the essential requirements for effective gender budgeting according to the provided information?**

   The essential requirements for effective gender budgeting include political will, gender-segregated data, and emphasizing transparency, partnership, and collaboration.

3. **How can political will contribute to the success of gender budgeting initiatives?**

   Political will involves a strong commitment from the political sphere to advance gender equality and recognize gender budgeting as a significant strategy. It ensures regular discussions on gender budgeting during budget meetings and promotes awareness in relevant political arenas.

4. **Why is gender-segregated data fundamental for gender budgeting?**

   Gender-segregated data serves as the foundation for evaluating the gender implications of policies. It helps in understanding the specific needs and priorities of different gender groups.

5. **What is the significance of transparency, partnership, and collaboration in gender budgeting?**

   Transparency is vital in the budgeting process and decision-making. Collaboration between budgetary experts and gender specialists, as well as the involvement of women and men at every stage, is essential. Collaboration extends to civil society organizations, external experts, and various stakeholders, fostering cooperation between government ministries and entities responsible for promoting gender equality.

6. **What are the goals and objectives of the Animal Husbandry Department concerning gender equality, and how do they align with the broader national goal of achieving gender equality?**

   The department's goals and objectives may include increasing women's participation in livestock-related activities, promoting women's ownership of livestock assets, and ensuring equitable access to veterinary services for both men and women. These align with the national goal of gender equality by empowering women in agriculture.

7. **How can the animal husbandry department identify the specific needs and priorities of women in animal husbandry?**

   Methods for identifying needs and priorities may include surveys, consultations, and data analysis to reveal women's needs, such as training, access to credit, and childcare support.

8. **Are the existing policies, plans, programs, and schemes in animal husbandry gender-sensitive and do they address women's specific needs?**

   An evaluation should be conducted to determine whether existing initiatives in animal husbandry consider gender-specific needs, such as gender-sensitive training and programs to encourage women's ownership of livestock.

9. **What activities can the Animal Husbandry Department undertake to reduce gender gaps in the sector?**

   The department can implement activities like gender-sensitive training, financial support for women in livestock ownership, and establishing women-friendly livestock service centres.

10. **What challenges does the department face in reaching women in animal husbandry, and how can these challenges be addressed?**

    Challenges may include limited mobility, traditional gender roles, and lack of awareness among women. Addressing these challenges could involve community-based training, awareness campaigns, and using local networks to disseminate information.

11. **What are five steps in the framework for gender budgeting?**

    The five steps in the framework for gender budgeting are:

    i. Analysis of the situation for women, men, girls, and boys in a given sector.

    ii. Assessment of the sector's policy addressing gender issues.

    iii. Evaluation of budget allocations for gender-sensitive policies.

    iv. Monitoring the budget's implementation.

    v. Assessment of the impact of policies on gender equality.

12. **In the context of gender budgeting, why is it important to assess the adequacy of budget allocations?**

    Assessing the adequacy of budget allocations is important to ensure that there is enough funding to support initiatives that promote gender equality. It helps determine whether the allocated resources are sufficient to implement gender-sensitive policies and programs effectively.

13. **How does spatial mapping contribute to gender-responsive programs, and what does it reveal?**

    Spatial mapping reveals deviations from norms and spatial variations in program implementation. It highlights areas that need special attention and can reveal disparities in resource distribution, access to services, and participation. It helps in tailoring programs to address specific geographical and gender-related challenges.

14. What is the significance of using gender-sensitive checklists in planning and budgeting?

Gender-sensitive checklists are crucial in planning and budgeting because they ensure that gender-specific considerations are integrated into program development. They help identify gender-specific activities, resource allocations, and performance evaluation metrics, thus promoting gender equity and inclusivity in policies and programs.

15. Give an example of a hypothetical gender-specific scheme in animal husbandry, and briefly describe the activities under it.

An example of a hypothetical gender-specific scheme in animal husbandry can be "Women Livestock Entrepreneurship Program." Activities under this scheme may include training women in livestock management practices, providing access to female-specific resources like training on dairy farming, poultry management, and animal healthcare.

16. What does the gender-sensitive checklist's performance audit stage aim to achieve?

The performance audit stage of the gender-sensitive checklist aims to review the actual performance of a program compared to its annual targets, identify constraints in achieving those targets, evaluate the impact of the program, and analyse trends in expenditure and impact indicators.

17. How can spatial mapping help in making corrective actions for gender-responsive programs?

Spatial mapping can identify areas with low participation and disparities in resource distribution. This information can guide corrective actions, such as allocating additional resources to underprivileged areas, conducting awareness campaigns, or improving infrastructure to ensure more inclusive and equitable outcomes.

18. Who is responsible for promoting gender budgeting at the central level in India, and which departments can be the nodal departments at the state level?

The Ministry of Women and Child Development is responsible at the central level, and at the state level, women and child development, social welfare, finance, or planning departments can be nodal departments.

19. What are the two parts of gender budgetary allocations in India, and can you provide examples of schemes in each part?

Part A includes schemes with 100% allocation for women, such as the widow pension scheme, girls' hostel scheme, and maternity benefit scheme. Part B includes schemes allocating 30% to 99% of funds for women, such as the mid-day meals program, the rural livelihoods mission, and the biogas program.

20. **What is the objective of the Female Calf Rearing Scheme (Sunandini), and how does it aim to achieve these objectives?**

The objectives are to increase the number of productive milch animals, achieve timely puberty and conception rates in crossbreed calves, and support rural poor to improve their livelihood through dairying. It aims to achieve these by providing calf feed, health, and insurance for female calves born through artificial insemination.

21. **Why does Gender Budgeting focus on women? Provide reasons mentioned in the paragraph.**

Gender Budgeting focuses on women due to reasons like women's high illiteracy rates, maternal mortality issues in developing nations, underrepresentation in decision-making roles, differences in economic contributions and pay, and their responsibilities for unpaid tasks like caregiving.

22. **What is the purpose of gender auditing in the context of gender budgeting, and when does it take place?**

Gender auditing in gender budgeting involves evaluating budget allocations and their impact on gender-related outcomes. It takes place after the budget's execution to ensure compliance and equitable benefits for women and men in the targeted sector, such as animal husbandry.

23. **How can gender budgeting in animal husbandry address gender disparities? Provide four strategies mentioned in the paragraph?**

Gender budgeting in animal husbandry can address gender disparities by:

a. Analyzing budget allocations for programs benefiting women.

b. Developing gender-sensitive indicators to measure budget impact.

c. Identifying areas with gender disparities and allocating funds to address them.

d. Ensuring marginalized groups like female-headed households are considered in budget allocations.

24. **What is the distinction between outputs and outcomes in program evaluation and why is it important?**

Outputs represent direct results of program activities, while outcomes are broader changes and benefits resulting from those activities. The distinction is important for evaluating program effectiveness and understanding the broader impact of interventions.

25. Provide an example of an output and an outcome in the context of animal husbandry.

Output: Measurable quantities of milk produced by individual cows due to program interventions.

Outcome: Broader impacts, such as increased income for farmers, improved local economy, and better community nutrition due to increased milk production.

26. Name two sectors traditionally considered gender-sensitive and two sectors traditionally considered gender-neutral.

Gender-sensitive sectors: Health and education.

Gender-neutral sectors: Defense and power.

27. What is gender mainstreaming, and what is its ultimate objective?

Gender mainstreaming is the process of evaluating the potential impacts on both women and men of any planned action, aiming to achieve gender parity as the ultimate objective.

28. Define gender-sensitive indicators and explain why they are important in promoting gender equality ?

Gender-sensitive indicators are measurements that aim to track changes for both women and men, covering quantitative and qualitative aspects. They are essential for promoting gender equality by assessing shifts in gender-related equity matters, collecting data disaggregated by sex, and capturing both quantitative and qualitative changes.

29. In the context of animal husbandry, provide a quantitative and a qualitative gender-sensitive indicator, and explain their importance.

Quantitative Indicator: Increase in Women's Ownership of Poultry Farms- This measures the percentage change in women-owned poultry farms, promoting economic empowerment & challenging traditional gender roles.

Qualitative Indicator: Shift in Gender Norms within Poultry Farming Communities - This measures changes in attitudes towards women's involvement in poultry farming, indicating progress toward more equitable gender relations.

30. What is the rationale behind the need for Gender budgeting in national budgets?

Gender Budgeting is necessary to address the distinct effects of budget allocation on males and females, as well as to reduce gender disparities in well-being, education, and economic prospects.

31. How does Gender Budgeting contribute to the achievement of gender equity and equality in India?

Gender Budgeting helps ensure that government commitments to gender parity and women's progression receive necessary financial support and mitigate unintentional impacts on gender disparities.

32. In what way can Gender Budgeting be a tool for monitoring the achievement of policy goals related to gender equality?

Gender Budgeting facilitates the gender-aware monitoring of policy goals, such as those outlined in the National Policy for Empowerment of Women 2001.

33. Why is it important to value unpaid work in the context of Gender Budgeting?

Unpaid work, such as childcare and household chores, contributes to a society's overall economic productivity. Recognizing and supporting individuals engaged in unpaid work, particularly women, reduces their burden and enhances social benefits.

34. How does Gender Budgeting address disparities in the animal husbandry sector, particularly in terms of resource access?

Gender Budgeting in animal husbandry aims to ensure equal access to resources like land, credit, and technology for both men and women engaged in dairy farming.

35. What are the implications of traditional gender roles in dairy farming for training and capacity building?

Traditional gender roles can lead to differences in the training received by men and women. Gender Budgeting allocates funds to provide tailored training programs to both genders in dairy farming.

36. How does Gender Budgeting promote women's economic empowerment in animal husbandry?

Gender Budgeting supports initiatives that help women add value to their dairy products and access broader markets, thereby increasing their income-generating potential.

37. What are some aspects of unpaid care work that Gender Budgeting in animal husbandry seeks to address?

Gender Budgeting in animal husbandry considers the time and effort women spend on tasks like fetching fodder, feeding animals, and cleaning, and allocates funds for mechanization or labor-saving technologies to reduce their workload.

38. What is the basis for implementing gender budgeting in national budgets?

Acknowledging that national budgets have distinct effects on males and females due to how resources are distributed.

39. Why is Gender Budgeting advocated as a means to foster gender mainstreaming?

Because it has the potential to transform gender imbalances caused by budget allocations and policies.

40. How does Gender Budgeting help monitor the achievement of policy goals in a gender-aware manner?

Gender Budgeting serves as a tool to monitor the achievement of policy goals, including the National Policy for Empowerment of Women, by ensuring government expenditures are aligned with gender perspectives.

41. Why is it important to value unpaid work in the context of Gender Budgeting?

Unpaid work, such as childcare and household chores, contributes to a society's overall economic productivity. Gender Budgeting aims to reduce the burden of unpaid work on both women and men by developing strategies and allocating resources accordingly.

42. In the context of animal husbandry, why is gender budgeting relevant?

Gender roles and responsibilities often influence resource allocation, access, and utilization in animal husbandry, making gender budgeting essential in addressing disparities and promoting gender equality.

43. How can Gender Budgeting address disparities in access to resources in dairy farming?

Gender budgeting can involve allocating funds to provide women with better access to resources such as land, training, and modern farming equipment to ensure equal opportunities in dairy farming.

44. What does Gender Budgeting aim to achieve regarding training and capacity building in dairy farming?

Gender budgeting aims to ensure that both men and women receive training programs tailored to their specific needs, so they can contribute effectively to the dairy farming enterprise.

45. How does Gender Budgeting support women in dairy farming to add value to their products and access broader markets?

Gender budgeting supports initiatives that help women in dairy farming add value to their products and access broader markets, thereby increasing their income-generating potential.

46. How can Gender Budgeting address the unequal burden of time and labour distribution in dairy farming?

Gender budgeting can allocate funds for mechanization or labor-saving technologies to reduce the workload on women who often bear the burden of unpaid care work in dairy farming.

SECTION D: OBJECTIVE QUESTIONS

Question 1. Multiple Choice Questions

1. What is the primary goal of Gender Budgeting?

a. Allocating funds exclusively for women

b. Promoting gender equality through budgetary measures

c. Funding projects for women's empowerment

d. Eliminating budget allocations for men

2. When did the concept of Gender Budgeting originate?

a. 1984 b. 1995

c. 2004 d. 2010

3. Which country was the first to implement Gender Budgeting?

a. India b. Australia

c. South Africa d. Uganda

4. What is the significance of the Women's Conference in Beijing in 1995 in relation to Gender Budgeting?

a. It marked the origin of Gender Budgeting.

b. It highlighted the need for gender-sensitive budget analysis.

c. It called for the complete elimination of gender-related budgets.

d. It focused on women's empowerment without budgetary considerations.

5. How many countries globally have embraced Gender Budgeting?

a. Over 50 b. Over 75

c. Over 100 d. Over 150

6. In India, when was Gender Budgeting introduced as a fiscal approach?

a. 1992 b. 1997

c. 2001 d. 2004-05

7. **What is the Women Component Plan (WCP) in India?**
   a. A plan to exclude women from budget considerations
   b. A plan to allocate funds for women's empowerment
   c. A plan to prioritize women's fashion and clothing
   d. A plan to support women's sports initiatives

8. **When did the momentum for gender-based budget analysis surge?**
   a. 1984 b. 1995
   c. 2004 d. 2010

9. **Which country took the lead in implementing Gender Budgeting after 1995?**
   a. Australia b. South Africa
   c. Switzerland d. China

10. **What is the main goal of gender budgeting?**
   a. To promote gender inequality
   b. To restructure budgets for economic growth
   c. To assess the impact of budgets on gender dynamics
   d. To achieve gender equality goals

11. **What is the role of Gender Budgeting Cells (GBCs) in Indian Ministries?**
   a. To exclude gender considerations from budgeting
   b. To ensure transparency in budget allocation
   c. To promote gender equality within the ministry
   d. To track men's and women's spending separately

12. **What is the Gender Budget Statement (GBS) in India?**
   a. A financial statement for gender-specific projects
   b. A statement excluding women from budget considerations
   c. A document highlighting gender disparities
   d. A component of the Union Budget focusing on gender issues

13. **What is the essential precondition for the success of gender budgeting initiatives?**
   a. Gender-segregated data b. Political will
   c. External experts d. Women's participation

14. **What does gender-segregated data provide the foundation for in gender budgeting?**

a. Transparency
b. Political will
c. Gender equality
d. Policy formulation

15. **In gender budgeting, what is the core principle guiding effective gender budgeting?**

a. Transparency, partnership, and collaboration
b. Gender-segregated data
c. Political will
d. Women's participation

16. **What does the process of gender budgeting necessitate collaboration between?**

a. Political parties
b. Gender budgeting cells
c. Budgetary experts and gender specialists
d. External experts and women's groups

17. **What does a gender-responsive budget aim to achieve?**

a. Excluding gender considerations from the budget
b. Allocating funds exclusively for women
c. Acknowledging gender patterns and changing them for gender equality
d. Eliminating funding for gender-specific projects

18. **What is the primary focus of gender budget initiatives?**

a. Excluding women from budget considerations
b. Promoting women's participation in politics
c. Moving the country towards a gender-responsive budget
d. Eliminating all budget allocations

19. **What does gender budgeting emphasize in resource allocation?**

a. Exclusively funding women's projects
b. Ignoring gender-specific needs
c. Promoting gender equality in budget allocation
d. Reducing the budget for non-gender-specific projects

20. What was the goal of the Eighth Five-Year Plan in India regarding gender-related allocations?

a. Allocating 30% of funds to women

b. Excluding women from budgetary considerations

c. Promoting gender-neutral budgeting

d. Eliminating all budget allocations for women

21. What does the National Policy for the Empowerment of Women in India aim to achieve?

a. Excluding women from policy considerations

b. Promoting gender-neutral policies

c. Empowering women through policy measures

d. Eliminating all policies related to women

22. Why is Gender Budgeting advocated in India?

a. To promote economic prosperity

b. To ensure gender parity and women's progress

c. To reduce overall government spending

d. To address disparities in technology access

23. What is one of the key objectives of Gender Budgeting?

a. Promoting gender stereotypes

b. Monitoring the achievement of policy goals

c. Ignoring unpaid work

d. Exacerbating gender disparities

24. How does Gender Budgeting address the issue of unpaid work?

a. By advocating for equal pay for equal work

b. By recognizing the importance of unpaid tasks in the economy

c. By eliminating unpaid work entirely

d. By increasing the burden on individuals engaged in unpaid work

25. Why is Gender Budgeting relevant in animal husbandry?

a. To promote gender stereotypes

b. To ensure women have no role in the sector

c. To address disparities in resource access and control

d. To make men solely responsible for dairy farming

26. What is an example of a disparity in animal husbandry ?

a. Men and women having equal access to resources

b. Women receiving the same training as men

c. Women being involved in larger-scale commercial dairy operations

d. Unequal access to land, credit, and technology

27. How does Gender Budgeting address time and labor distribution in animal husbandry?

a. By increasing women's workload

b. By providing women with more unpaid care work

c. By allocating funds for mechanization and labor-saving technologies

d. By emphasizing traditional gender roles

28. What is the purpose of gender auditing in the context of animal husbandry budgeting?

a. To allocate funds for animal husbandry programs

b. To evaluate the outcomes of budget execution from a gender perspective

c. To create gender-neutral budget allocations

d. To identify areas where gender disparities exist

29. In the distinction between outputs and outcomes, which of the following represents an output in the context of an animal husbandry program aiming to increase milk production?

a. Increased income for local dairy farmers

b. Improved community nutrition

c. Measurable quantities of milk produced by individual cows

d. Broader impacts on the local economy

30. Which of the following sectors is considered gender-sensitive due to clearly identifiable beneficiaries, primarily women?

a. Defense b. Power

c. Health d. Trade

31. What is the ultimate objective of gender mainstreaming?

a. To create gender-neutral policies

b. To achieve parity between genders

c. To exclude men from the decision-making process

d. To focus exclusively on women's issues

32. What do gender-sensitive indicators aim to measure in the context of animal husbandry?

a. Quantitative changes for men only

b. Qualitative changes in women's attitudes

c. Relative changes for both women and men

d. Changes in livestock production

33. In the context of animal husbandry, which of the following is a quantitative gender-sensitive indicator for improving women's participation in poultry farming?

a. Increase in women's ownership of poultry farms

b. Empowerment of women in poultry decision-making

c. Shift in gender norms within poultry farming communities

d. Changes in poultry market prices

34. Why is measuring the shift in gender norms within poultry farming communities important in the context of gender-sensitive indicators?

a. It helps increase poultry production.

b. It reflects changes in community income.

c. It indicates progress toward more equitable gender relations.

d. It quantifies the number of women involved in poultry farming.

35. Why does Gender Budgeting tend to focus on women?

a. To promote gender equality in all aspects of society

b. Because women make up nearly two-thirds of the global illiterate population

c. To address issues like maternal mortality and gender-based violence

d. All of the above

36. In the context of Gender Budgeting, what is the significance of women's underrepresentation in decision-making roles?

a. It indicates a lack of interest among women in leadership positions.

b. It highlights the need for equal representation of women in government and corporate leadership.

c. It is not relevant to Gender Budgeting efforts.
d. It reflects women's preference for lower-status employment.

37. What is one of the key factors contributing to unequal pay for women in the workforce?

a. Women's lack of education and skills
b. Women's preference for informal employment
c. Differences in women's economic contributions compared to men
d. Gender-based discrimination and bias

38. Why is it important to address unpaid tasks such as childbirth, upbringing, and caregiving in Gender Budgeting efforts?

a. These tasks are primarily the responsibility of men, not women.
b. Addressing these tasks can help women achieve higher-paying jobs.
c. Unpaid tasks can limit women's participation in the formal workforce and affect their economic well-being.
d. Unpaid tasks are not a significant concern in Gender Budgeting.

39. Which part of the Gender budgetary allocations in India includes schemes allocating 30% to 99% of funds for women?

a. Part A
b. Part B
c. Part C
d. Part D

40. Which ministry or department at the central level is responsible for promoting gender budgeting in India?

a. Ministry of Finance
b. Ministry of Agriculture
c. Ministry of Women and Child Development
d. Ministry of Health

41. What is the purpose of the Gender Budget Cell?

a. To implement gender-specific programs
b. To facilitate gender analysis in government budgets
c. To conduct gender audits
d. To oversee social welfare programs

42. What is the primary focus of gender budgeting?

a. Promoting government efficiency
b. Addressing gender imbalances and promoting gender equality

c. Reducing government expenditures

d. Increasing tax revenues

43. What does spatial mapping help identify?

a. Geographic locations for new government offices

b. Troubled areas for government intervention

c. Areas with low population density

d. Gender-related issues in program implementation

44. How many ministries and departments at the government of India level have set up gender budgeting cells to pursue gender budgeting?

a. 5 b. 15

c. 57 d. 100

45. What is the primary role of gender-segregated data in gender budgeting?

a. To assess political will

b. To evaluate the gender implications of policies

c. To promote government commitment

d. To raise awareness in the political sphere

46. Who should be involved in the collaborative approach to gender budgeting?

a. Only government ministries

b. Only budgetary experts

c. Both women and men

d. Both external experts and civil society organizations

47. Which of the following is not a question that a department or organization should address during gender budgeting?

a. Identifying the needs and priorities of women

b. Assessing the challenges in reaching women

c. Setting goals and objectives for gender equality

d. Analyzing the economic impact of budgeting

48. What is one of the challenges that the Animal Husbandry sector might face in reaching women?

a. Lack of awareness about available schemes and programs

b. Adequate training for women in livestock management

c. Support for women's ownership of livestock
d. Access to veterinary services for both men and women

49. The movement for gender-based budget analysis exploded following:

a. The fourth Women's Conference in 1995 in Beijing
b. The first Women's Conference in 1985 in Kenya
c. The fifth Women's Conference in 1985 in Tokyo
d. The third Women's Conference in 1975 in New Delhi

50. In ...., the Indian government first incorporated gender budgeting into its fiscalstrategy.

a. 2002–2003
b. 2003–2004
c. 2004–2005
d. 2005–2006

51. Women Component Plan as a key strategy:

a. The Seventh Five-Year Plan
b. The Eighth Five-Year Plan
c. The Ninth Five-Year Plan
d. The Tenth Five-Year Plan

52. The National Policy for the Empowerment of Women was adopted by in2001:

a. National Commission for Women
b. The Committee for Women and Child Development
c. The Ministry of Women and Child Development
d. None of the above

53. All Ministries were mandated to establish .... in 2005:

a. Gender Budgeting Committees (GBCs)
b. Gender Budgeting Centers (GBCs)
c. Gender Budgeting Cells (GBCs)
d. Gender Budgeting Chains (GBCs)

54. The GBS was first included in the Union Budget in 2005–06, beginning an annualtradition. It stands for?

a. Gender Budget Sanction
b. Gender Budget Statement
c. Gender Budget Section
d. Gender Budget Statistic

55. Which one is not required for Gender Budgeting:

a. Political will
b. Gender-segregated data

c. Transparency, partnership and collaboration

d. Emphasizing Social Reforms

56. Trouble spots that need special attention are highlighted by:

a. Five step Framework

b. Participatory planning and budgeting

c. Gender Sensitive Checklists

d. Spatial Mapping

57. One of the challenges in access of training and services to livestock rearing women inrural areas:

a. Limited Mobility

b. Lack of Awareness

c. Poor sex-ratio

d. Shortage of logistics support

58. Tool(s) for Deepening gender budgeting:

a. Five step Framework for Gender budgeting

b. Participatory planning and budgeting

c. Spatial Mapping and Gender Sensitive Checklists

d. All above

59. Impact assessment in the five-step framework for Gender Budgeting has nothing to dowith:

a. Any change in income levels for both women and men.

b. Any change in ability of women to take on more diverse roles.

c. Decision-making becoming more/less inclusive?

d. Examining whether there is enough funding to support initiatives that promote gender equality.

60. How many ministries and departments in GOI have setup gender budgeting cells topursue gender budgeting ?

a. 37

b. 47

c. 57

d. 67

61. Part A of the gender budgetary allocations in India includes schemes for women suchas

a. The mid-day meals programme

b. The rural livelihoods mission

c. The widow pension scheme

d. The biogas programme

62. Part B of the gender budgetary allocations in India includes schemes for women, suchas

a. The mid-day meals programme

b. The rural livelihoods mission

c. The biogas programme

d. All above

63. Scrutinizing financial allocations, observing trends over a period, evaluating percentagedistributions and comprehensively analyzing the established systems, employed processes, as well as the results and effects of budget allocations, all from a gender perspective occurs in:

a. Gender Sanctioning

b. Gender Auditing

c. Gender Formulation

d. Gender Systematization

64. The quantifiable measure of goods and services generated through an activity within a scheme or program.

a. Outcome

b. Outreach

c. Outputs

d. Benefit

65. Final results produced by various governmental initiatives and interventions may be considered as their:

a. Outcome

b. Outreach

c. Outputs

d. Benefit

66. What is the main purpose of integrating gender budgeting into animal husbandry practices?

a) Increasing the cost of animal husbandry operations

b) Promoting gender-specific animal breeds

c) Enhancing equity and inclusivity in the sector

d) Focusing solely on women's empowerment

67. How can gender budgeting improve access to veterinary services for women and men in rural areas?

a. By excluding men from receiving veterinary care

b. By establishing mobile veterinary clinics

c. By reducing access to healthcare services

d. By focusing only on urban areas

68. What could gender budgeting allocate funds for in order to support women's financial inclusion in animal husbandry?

a. Microfinance programs tailored to women's needs
b. Large-scale agricultural enterprises
c. Programs exclusively for men's financial empowerment
d. Luxury livestock breeds

69. What is the purpose of Step 1 in the Five-Step Framework for Gender Budgeting?

a. Assess the impact of gender-sensitive policies
b. Review existing policies and regulations
c. Examine the budget allocation
d. Analyze roles and responsibilities within the sector

70. Which step involves tracking whether the allocated budget was used as planned?

a. Step 1: Analysis of the Situation
b. Step 2: Assessment of Sector Policy
c. Step 3: Adequacy of Budget Allocations
d. Step 4: Monitoring Implementation

71. What is the purpose of Step 5 in the framework?

a. Review existing policies and regulations
b. Assess the impact of gender-sensitive policies and programs
c. Examine the budget allocation
d. Analyze roles and responsibilities within the sector

72. What is the purpose of a Gender Budget Cell?

a. To allocate funds exclusively for women's projects
b. To integrate gender analysis into government budgets
c. To audit government spending
d. To oversee civil society organizations

73. When was the Gender Budget Statement introduced in the Union Budget of India?

a. 1995-96 b. 2000-01
c. 2005-06 d. 2010-11

74. What is the purpose of gender auditing in the context of gender budgeting?

a. To allocate funds specifically to women's programs
b. To track the increase in men's participation in various sectors
c. To analyze the outcomes and effects of budget allocations from a gender perspective
d. To assess the overall financial health of the budget

75. Which of the following is an example of an "outcome" in gender budgeting for animal husbandry?

a. The number of women trained in livestock farming techniques
b. The amount of funds allocated for veterinary services
c. An increase in the income of local dairy farmers due to higher milk yields
d. The percentage of women-headed households in rural areas

76. Gender budgeting is primarily concerned with:

a. Allocating more funds to women's programs than men's programs
b. Ensuring equal distribution of resources between genders
c. Analyzing the budget's impact on gender-related outcomes
d. Excluding men from budgetary processes

Question 2. True or False

1. Gender Budgeting aims to address gender disparities and promote gender equality in various sectors, including animal husbandry.
2. Gender Budgeting only focuses on issues directly related to human beings and does not have relevance in sectors like animal husbandry.
3. Gender Budgeting involves assessing and addressing disparities in access to resources, training, marketing, and time distribution in various sectors.
4. Women might have unequal opportunities in dairy farming due to traditional gender norms, leading to disparities in access to resources.
5. Gender budgeting in animal husbandry can help increase the income-generating potential of women by supporting initiatives that add value to their dairy products and access broader markets.
6. Gender Budgeting seeks to reduce the workload of women in unpaid care work by allocating funds for mechanization or labor-saving technologies.

7. Gender auditing in animal husbandry is conducted before the budget's execution.
8. Gender-sensitive indicators in animal husbandry can include both quantitative and qualitative measurements.
9. Gender mainstreaming aims to achieve gender parity in all sectors, including traditionally gender-sensitive ones.
10. An increase in women's ownership of poultry farms is a qualitative indicator in animal husbandry.
11. Gender-sensitive indicators in animal husbandry focus solely on quantitative aspects, such as income levels and literacy rates.
12. Gender budgeting primarily focuses on women.
13. Women's literacy rates are consistently higher than men's globally.
14. Maternal mortality is a significant cause of death for women in their reproductive years in developing nations.
15. Women are well-represented in decision-making roles in both governmental and corporate spheres.
16. Women's economic contributions are often characterized by informal and lower-status employment, leading to equal pay for equivalent work.
17. Women are primarily responsible for unpaid tasks such as childbirth, upbringing, and caregiving for dependents.
18. In India, the Ministry of Women and Child Development is responsible for promoting gender budgeting at both the central and state levels.
19. The institutional mechanism for integrating gender analysis into the government budget in India is called the Gender Budget Cell.
20. Gender budgeting focuses on addressing gender imbalances and promoting gender equality within government budgets.
21. Spatial mapping is used to identify areas where gender-related issues in program implementation need special attention.
22. The Five-Step Framework for Gender Budgeting includes steps for analyzing the impact of policies and programs on gender equality.
23. Political will is not a crucial factor in the success of gender budgeting initiatives.
24. Gender-segregated data is not essential for evaluating the gender implications of policies in gender budgeting.
25. Transparency, partnership, and collaboration are not core principles in effective gender budgeting.

26. Collaboration in gender budgeting is limited to government ministries and does not involve external experts or civil society organizations.
27. The Committee on the Status of Women's report in India was released in 1974.
28. Gender Budgeting is only focused on the allocation of resources and doesn't cover policies and programs.
29. Gender Budgeting was integrated into India's fiscal strategy in 2007-08.
30. Gender budget initiatives aim to move the country towards a less gender-responsive budget.
31. The concept of gender budgeting was first introduced in Australia in 1984.
32. Gender budgeting gained momentum after the fourth Women's Conference in Beijing in 1995.
33. Over 100 countries around the world have adopted gender budgeting.
34. The Government of India integrated Gender Budgeting into its fiscal strategy in the 2004-05 fiscal year.
35. The Ninth Five-Year Plan in India emphasized allocating "not less than 30 percent of funds/benefits" in all sectors related to women.
36. The Gender Budget Statement (GBS) was introduced in the Indian Union Budget in the 2005-06 fiscal year.
37. The concept of gender budgeting primarily focuses on restructuring expenditures to exclusively benefit women.
38. Gender budgeting involves assessing the impact of budgets from a gender perspective and restructuring resources to promote gender equality.
39. Gender budgeting is solely concerned with financial allocations and does not involve policy formulation or impact assessment.
40. Gender budgeting aims to create a budget that acknowledges and addresses gender patterns in society to move towards a more gender equal society.
41. The AH department needs to evaluate whether outdated policies, plans, programs and schemes in animal husbandry address gender-specific needs.
42. For deepening gender budgeting, mere allocation of money is sufficient and only criteria to accomplish.
43. Gender budgeting takes place subsequent to the Gender auditing.
44. Outputs provide a measure of the direct results of activities, outcomes focus on the broader changes and benefits that result from those activities.

45. Gender budgeting in animal husbandry could involve allocating resources for training sessions that accommodate different schedules, benefiting both men and women involved in the sector.
46. Gender budgeting may prioritize the construction of infrastructure like water points or feed storage facilities in locations that are convenient for men, as they are the primary caregivers for livestock.
47. Gender-disaggregated data collection is not necessary for effective gender budgeting in animal husbandry practices.
48. Gender budgeting only focuses on women because men do not face any gender-related disparities.
49. The Gender Budget Cell is responsible for exclusively allocating funds to women's programs.
50. Gender auditing takes place before the execution of the budget to ensure compliance with gender equality goals.
51. Outputs refer to the final results and impacts of governmental initiatives.
52. Gender budgeting only focuses on women's needs and programs, excluding men's concerns.
53. Gender-neutral sectors are those in which revenue and expenses can be easily divided based on gender.
54. Gender-sensitive indicators only focus on quantitative changes and don't consider qualitative aspects.
55. Gender budgeting aims to integrate gender perspectives into the entire budgetary process.
56. Gender budgeting is only relevant to issues directly related to human beings and not to sectors like animal husbandry.
57. Gender budgeting in animal husbandry can help address disparities in resource access, training, marketing, and time distribution.
58. Gender mainstreaming involves evaluating the potential impacts on women only, without considering men.
59. Gender-sensitive indicators in animal husbandry can measure changes in social norms but not changes in women's participation.
60. Gender-sensitive sectors are those where the beneficiaries, primarily men, are clearly identifiable.
61. Gender budgeting is a tool to monitor the achievement of the goals of the National Policy for Empowerment of Men.

62. Traditional gender norms do not influence the allocation of resources and responsibilities in sectors like animal husbandry.
63. Gender budgeting aims to reinforce existing gender disparities within the Animal Husbandry sector.
64. Political will is not necessary for the success of gender budgeting initiatives.
65. Gender-segregated data is not important for gender budgeting.
66. Collaboration between budgetary experts and gender specialists is not necessary for gender budgeting.
67. The inclusion of external experts is not beneficial for gender budgeting.
68. Gender budgeting only involves cooperation between the Ministry of Finance and gender equality entities.
69. Gender budgeting does not require assessing the challenges faced in reaching women and girls with services.
70. Gender budgeting does not require aligning Ministry/Department's goals with the larger National Goal of achieving Gender Equality.
71. Gender budgeting in the Animal Husbandry sector focuses solely on women's participation in livestock management.
72. Gender budgeting aims to allocate funds exclusively to women's initiatives within the Animal Husbandry sector.
73. Gender budgeting is a standalone process that doesn't need to align with broader national goals.
74. Gender budgeting in Animal Husbandry focuses solely on financial allocations.
75. Gender budgeting is concerned only with addressing challenges faced by women, neglecting issues faced by men.
76. Gender budgeting in Animal Husbandry only involves allocating funds for women's training programs.

Question 3. Fill in the Blanks

1. Gender budgeting was first initiated in __________ in 1984.
2. In 2004-05, the Government of __________ integrated Gender Budgeting into its fiscal strategy.
3. The Women Component Plan required both Central and State Governments to allocate "not less than ______ percent of funds/benefits" in all sectors related to women.

4. Gender Budgeting Cells (GBCs) within all Ministries were mandated in __________.
5. The introduction of the Gender Budget Statement (GBS) in the Union Budget marked an annual practice in __________.
6. Gender budgeting is a tool that seeks to integrate _______ perspectives into the entire budgetary process.
7. Gender budgeting aims to address disparities in access to resources, such as _______ and technology, in the field of animal husbandry.
8. Gender budgeting considers the time and effort spent by women on unpaid care work, such as fetching fodder, feeding animals, and cleaning. Funds are allocated for mechanization or labour-saving technologies to reduce their _______.
9. Gender auditing serves as a compliance measure for ________.
10. In gender budgeting for animal husbandry, a financial review during gender auditing assesses how allocated funds were spent on ________ programs.
11. The ultimate objective of gender mainstreaming is to achieve ________ between genders.
12. Gender-sensitive sectors are traditionally those in which beneficiaries, primarily ________, are clearly identifiable.
13. Gender-sensitive indicators aim to measure relative changes for both women and men and contribute to promoting ________.
14. In the context of animal husbandry, a quantitative gender-sensitive indicator is the increase in ___ participation in poultry training programs.
15. The qualitative gender-sensitive indicator that assesses changes in attitudes toward women's roles in animal husbandry is the shift in ________ within poultry farming communities.
16. Gender auditing involves a comprehensive analysis of budget allocations from a ________ perspective.
17. Outputs in animal husbandry represent quantifiable measures of goods and services generated through ________ activities.
18. An outcome in animal husbandry might be an increase in the income of local dairy farmers due to ________ milk yields.
19. Gender Budgeting tends to focus on women because women make up nearly _______ of the global illiterate population.
20. In developing nations, maternal mortality remains a significant cause of death for _______ in their reproductive years.

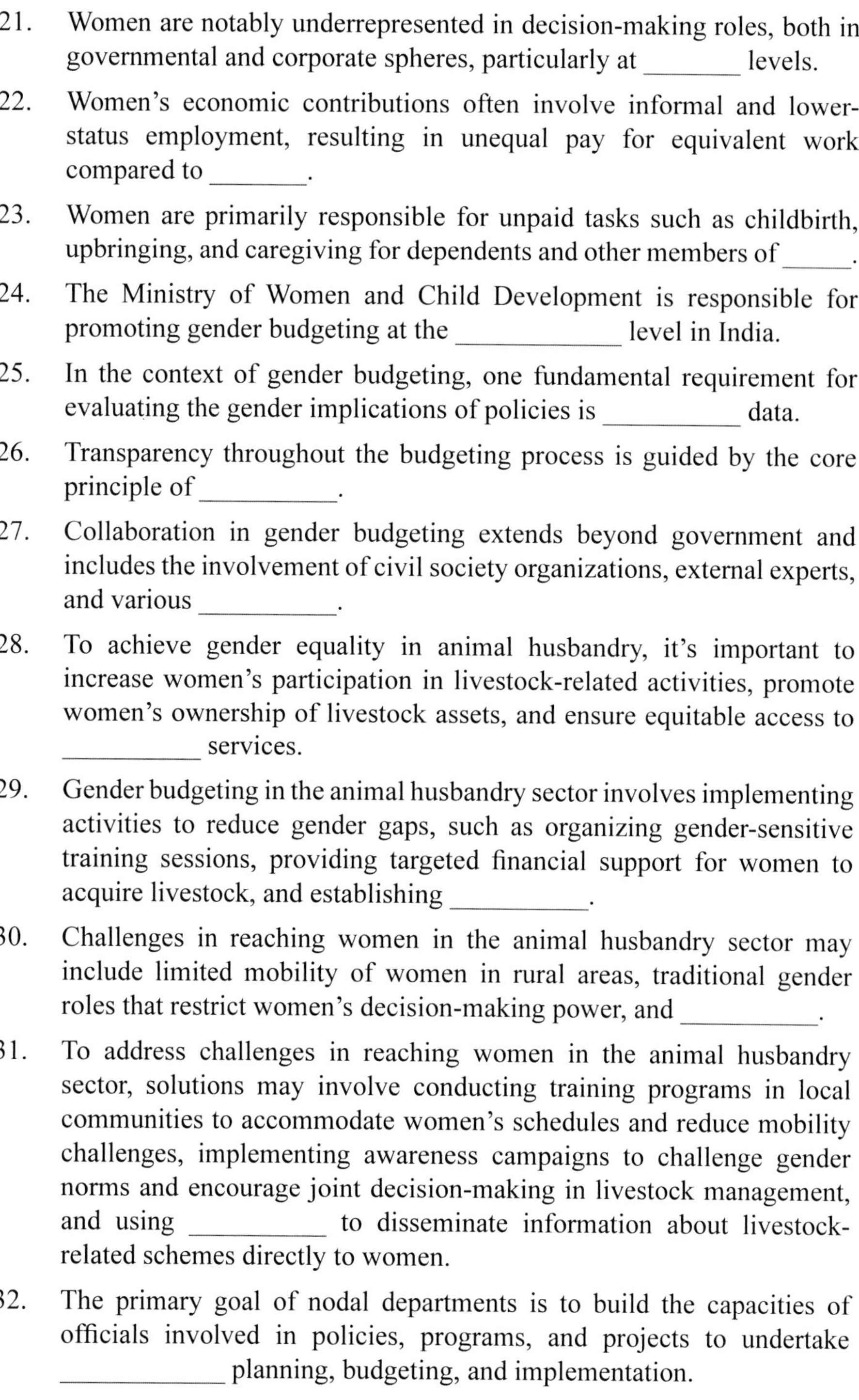

21. Women are notably underrepresented in decision-making roles, both in governmental and corporate spheres, particularly at _______ levels.
22. Women's economic contributions often involve informal and lower-status employment, resulting in unequal pay for equivalent work compared to _______.
23. Women are primarily responsible for unpaid tasks such as childbirth, upbringing, and caregiving for dependents and other members of _____.
24. The Ministry of Women and Child Development is responsible for promoting gender budgeting at the ___________ level in India.
25. In the context of gender budgeting, one fundamental requirement for evaluating the gender implications of policies is __________ data.
26. Transparency throughout the budgeting process is guided by the core principle of __________.
27. Collaboration in gender budgeting extends beyond government and includes the involvement of civil society organizations, external experts, and various __________.
28. To achieve gender equality in animal husbandry, it's important to increase women's participation in livestock-related activities, promote women's ownership of livestock assets, and ensure equitable access to __________ services.
29. Gender budgeting in the animal husbandry sector involves implementing activities to reduce gender gaps, such as organizing gender-sensitive training sessions, providing targeted financial support for women to acquire livestock, and establishing __________.
30. Challenges in reaching women in the animal husbandry sector may include limited mobility of women in rural areas, traditional gender roles that restrict women's decision-making power, and __________.
31. To address challenges in reaching women in the animal husbandry sector, solutions may involve conducting training programs in local communities to accommodate women's schedules and reduce mobility challenges, implementing awareness campaigns to challenge gender norms and encourage joint decision-making in livestock management, and using __________ to disseminate information about livestock-related schemes directly to women.
32. The primary goal of nodal departments is to build the capacities of officials involved in policies, programs, and projects to undertake ___________ planning, budgeting, and implementation.

33. The Ministry of Finance, both at the central and state levels, plays a crucial role in the ____________ of gender budgeting.
34. ____________ Departments, such as the Ministry of Agriculture, Health, Education, and Labor, are involved in gender budgeting efforts to ensure sectoral inclusivity.
35. The ____________ and Audit General of India, along with local audit departments, are responsible for monitoring and evaluating the effectiveness of gender budgeting.
36. Gender budgeting involves the integration of gender analysis into the government ____________, with the aim of addressing gender imbalances and promoting gender equality.
37. Civil society organizations, researchers, and ____________ are considered important stakeholders in the gender budgeting process.
38. In __________ (country), the first gender budgeting initiative was launched in 1984.
39. The fourth Women's Conference in 1995 was held in _______ (country).
40. Since 2004–2005, when the Indian government incorporated it into its fiscal strategy; ___________ has been a regular part of India's annual budget.
41. Both the Central and State Governments were directed to allocate "not less than _______ percent of funds/benefits" in all sectors related to women.
42. Gender budgeting involves allocating resources in a way that advances gender ____________.
43. Women have to be treated as equal partners in ___________ and _________ of planning and budgeting rather than only as beneficiaries.
44. The Gender budgetary allocation in India has________ parts.
45. Part A of the gender budgetary an allocation in India includes schemes with _____ allocation while Part B allocating _____ of funds for women.
46. Gender ______ functions as a compliance measure for gender budgeting.
47. Gender budgeting aims to ensure __________ and inclusivity by considering the different needs and priorities of men and women in animal husbandry.
48. One way gender budgeting can enhance women's economic empowerment is by providing access to tailored __________ programs for investing in livestock.

49. Allocating resources for market linkages and networking opportunities can improve __________ access for both men and women involved in animal husbandry.
50. Gender-disaggregated __________ can provide insights into the unique challenges faced by women and men in the animal husbandry sector.
51. The Ministry of _________ and Child Development is responsible for promoting gender budgeting in India at the central level.
52. A Gender Budget Cell facilitates the integration of _________ analysis into government budgets to promote gender equality.
53. The Gender Budget Statement was introduced in the Union Budget of India in the fiscal year _________.
54. Part A of the Gender Budget includes schemes with _________ allocation for women.
55. Gender budgeting focuses on women due to reasons such as unequal _________, underrepresentation, and caregiving responsibilities.
56. Gender auditing involves scrutinizing financial allocations and evaluating the ____ and effects of budget allocations from a gender perspective.
57. An increase in the income of local dairy farmers due to higher milk yields is an example of an _________ in animal husbandry.
58. Gender budgeting aims to ensure that both women and men benefit _________ from budget allocations.
59. Gender-sensitive indicators aim to measure relative changes for ________ and ________.
60. Gender mainstreaming involves evaluating the potential impacts on both ________ and ________ of any planned action.
61. Gender budgeting aims to address gender disparities and promote gender ________.
62. One example of a quantitative indicator in animal husbandry is the ________ in Women's Ownership of Poultry Farms.
63. Gender-sensitive indicators not only focus on quantitative changes but also on the ________ that lead to empowerment.
64. Gender budgeting ensures that government expenditures and policies are designed to address ________ disparities.
65. Gender budgeting can play a crucial role in sectors like ____ husbandry.
66. In many societies, women often have limited ________ and control over productive resources.

67. Gender budgeting can help allocate funds for ________ or labor-saving technologies to reduce women's workload.

68. Gender-sensitive indicators aim to promote gender equality and gauge shifts in gender-related ________ matters

ANSWERS

1. Multiple Choice Questions

1	b	Promoting gender equality through budgetary measures
2	a	1984
3	b	Australia
4	b	It highlighted the need for gender-sensitive budget analysis
5	c	Over 100
6	d	2004-05
7	b	A plan to allocate funds for women's empowerment
8	b	1995
9	b	South Africa
10	d	To achieve gender equality goals
11	c	To promote gender equality within the ministry
12	d	A component of the Union Budget focusing on gender issues
13	b	Political will
14	a	Transparency
15	a	Transparency, partnership, and collaboration
16	c	Budgetary experts and gender specialists
17	c	Acknowledging gender patterns and changing them for gender equality
18	c	Moving the country towards a gender-responsive budget
19	c	Promoting gender equality in budget allocation
20	a	Allocating 30% of funds to women
21	c	Empowering women through policy measures
22	b	To ensure gender parity and women's progress
23	b	Monitoring the achievement of policy goals
24	b	By recognizing the importance of unpaid tasks in the economy
25	c	To address disparities in resource access and control
26	d	Unequal access to land, credit, and technology

27	c	By allocating funds for mechanization and labor-saving technologies
28	b	To evaluate the outcomes of budget execution from a gender perspective
29	c	Measurable quantities of milk produced by individual cows
30	c	Health
31	b	To achieve parity between genders
32	c	Relative changes for both women and men
33	a	Increase in women's ownership of poultry farms
34	c	It indicates progress toward more equitable gender relations.
35	d	All of the above
36	b	It highlights the need for equal representation of women in government and corporate leadership.
37	d	Gender-based discrimination and bias
38	c	Unpaid tasks can limit women's participation in the formal workforce and affect their economic well-being
39	b	Part B
40	c	Ministry of Women and Child Development
41	b	To facilitate gender analysis in government budgets
42	b	Addressing gender imbalances and promoting gender equality
43	d	Gender-related issues in program implementation
44	c	57
45	b	To evaluate the gender implications of policies
46	d	Both external experts and civil society organizations
47	d	Analyzing the economic impact of budgeting
48	a	Lack of awareness about available schemes and programs
49	a	The fourth Women's Conference in 1995 in Beijing
50	c	2004–2005
51	c	The Ninth Five-Year Plan
52	c	The Ministry of Women and Child Development
53	c	Gender Budgeting Cells (GBCs)
54	b	Gender Budget Statement
55	d	Emphasizing Social Reforms
56	c	Spatial Mapping

57	a	Limited Mobility
58	d	All above
59	d	Examining whether there is enough funding to support initiatives that promote gender equality.
60	c	57
61	c	The widow pension scheme
62	d	All above
63	b	Gender Auditing
64	c	Outputs
65	a	Outcome
66	c	Enhancing equity and inclusivity in the sector
67	b	By establishing mobile veterinary clinics
68	a	Microfinance programs tailored to women's needs
69	d	Analyze roles and responsibilities within the sector
70	d	Step 4: Monitoring Implementation
71	b	Assess the impact of gender-sensitive policies and programs
72	b	To integrate gender analysis into government budgets
73	c	2005-06
74	c	Toanalyze the outcomes and effects of budget allocations from a gender perspective
75	c	An increase in the income of local dairy farmers due to higher milk yields
76	c	Analyzing the budget's impact on gender-related outcomes

Answer 2. True and False

S.No.	True or false	S.No.	True or false
1	True	2	False
3	False	4	True
5	True	6	True
7	False	8	True
9	True	10	False
11	False	12	False
13	False	14	True
15	False	16	True
17	True	18	False

19	True	20	True
21	True	22	True
23	False	24	False
25	False	26	False
27	True	28	False
29	False	30	False
31	True	32	True
33	True	34	True
35	True	36	True
37	False	38	True
39	False	40	True
41	True	42	False
43	False	44	False
45	False	46	False
47	False	48	True
49	False	50	False
51	False	52	False
53	False	54	False
55	False	56	False
57	False	58	False
59	False	60	False
61	False	62	False
63	False	64	False
65	False	66	False
67	False	68	False
69	False	70	False
71	False	72	False
73	False	74	False
75	False	76	False

Answer 3. Fill in the Blanks

S.No.	Answer
1	Australia
2	India
3	30
4	2005

5	2005-06
6	Gender
7	Land
8	Workload
9	Gender budgeting
10	Gender-specific
11	Parity
12	Women
13	Gender equality
14	Women's
15	Gender norms
16	Gender
17	Program
18	Higher
19	Two thirds
20	Women
21	Senior
22	Men
23	Society
24	Central
25	Gender-segregated
26	Transparency
27	Stakeholders
28	Veterinary
29	Women-friendly livestock service centers
30	Lack of awareness among women about available schemes and programs
31	Community-based networks
32	Gender-sensitive
33	Implementation
34	Sectoral Ministries
35	Comptroller
36	Budget
37	Economists and Statistician

38	Australia
39	Beijing
40	Gender budgeting
41	30
42	Equality
43	Decision making, implementation
44	Two
45	100%, 30% to 99%
46	Auditing
47	Equity
48	Microfinance
49	Market
50	Data
51	Women
52	Gender
53	2005-06
54	100%
55	Economic contributions
56	Results
57	Outcome
58	Equitably
59	Women and men
60	Women and men
61	Equality
62	Increase
63	Processes
64	Gender
65	Animal
66	Ownership
67	Mechanization
68	Equity

Bibliography

Access Development Services. (2011). Catalysing Markets through Collectives: Experiences from the Allied Sector - Sitaram Rao Livelihoods India Case Study Competition. Edge Communications, New Delhi.

Aksornkool, A., Joerger, N., & Cindy, K. (2008). Gender Sensitivity: A Training Manual for Sensitizing Education Managers, Curriculum and Material Developers and Media Professionals to Gender Concerns. Washington, DC.

American Psychological Association. (2010). Publication Manual of the American Psychological Association (6th ed.). Washington, DC.

Asian Development Bank. (2013). Gender Equality and Food Security: Women's Empowerment as a Tool against Hunger. Philippines. ISBN 978-92-9254-171-2.

Assan, N. (2014). Gender Disparities in Livestock Production and Their Implication for Livestock Productivity in Africa. Scientific Journal of Animal Science, 3(5), 126-138.

Bain, C., Ransom, E., & Halimatusa'diyah, I. (2018). Weak Winners of Women's Empowerment: The Gendered Effects of Dairy Livestock Assets on Time Poverty in Uganda. Journal of Rural Studies, 61, 100–109.

Bishop-Sambrook, C., & Puskur, R. (2007). Toolkit for Gender Analysis of Crop and Livestock Production, Technologies, and Service Provision. International Livestock Research Institute ILRI, Nairobi (Kenya).

Blum, M. L., Cofini, F., & Sulaiman, R. V. (2020). Agricultural Extension in Transition Worldwide: Policies and Strategies for Reform. FAO, Rome.

Bravo-Baumann, H. (2000). Livestock and Gender: A Winning Pair: Capitalisation of Experiences on the Contribution of Livestock Projects to Gender Issues. Working Document, Swiss Development Cooperation, Bern.

Budlender, D., & Sharp, R. (1998). How to Do a Gender-Sensitive Budget: Contemporary Research and Practice. Commonwealth Secretariat and AusAid, London.

CBGA (2021). Training Resource on Gender Responsive Budgeting in India. Centre for Budget and Governance Accountability, New Delhi,India.

CGIAR. (2018). Making the Case for Gender: Livestock. Retrieved from https://cgspace.cgiar.org/server/api/core/bitstreams/35821687-da5b-46ce-8ae9-4c21483c5568/content

Chakravarty, R., & Dhaka, J. P. (1995). Delving into Gender Analysis. In K. Singh & J. B. Schiere (Eds.), Handbook for Straw Feeding Systems (pp. 20-25). ICAR, New Delhi, India.

Chander, M. (2023). Women-inclusive Livestock Development Helps Improve Women's Empowerment. Agrilinks. Retrieved from https://agrilinks.org/post/women-inclusive-livestock-development-helps-improve-womens-empowerment

Chauhan, Dharmistha. (2021). Training Manual on Gender and Climate Change Resilience. Kuala Lumpur and Bangkok: The Asian-Pacific Resource and Research Centre for Women (ARROW) and UN Women Regional Office for Asia and the Pacific.

Chauhan, T., & Jaffrelot, C. (2023). In Politics and Bureaucracy, Women Are Severely under-Represented. The Indian Express. Retrieved from https://indianexpress.com/article/opinion/columns/in-politics-and-bureaucracy-women-are-severely-under-represented-8492805/

Christoplos, I. (2010). Mobilizing the Potential of Rural and Agricultural Extension. Rome, FAO. Retrieved from http://www.fao.org/docrep/012/i1444e/i1444e00.pdf

CJI. (2023). Handbook on Combating Gender Stereotypes. Chief Justice of India, Supreme Court. India.

Colverson, K. E. (2015). Integrating Gender into Rural Advisory Services. GFRAS Good Practice Note 4. Lindau, Switzerland: GFRAS. Retrieved from www.g-fras.org/en/download.html?download=345:ggp-note-4-integrating-gender-into-rural-advisory-services

Council of Europe. (2005). Gender Budgeting: Final Report of the Group of Specialists on Gender Budgeting (EG-S-GB). Equality Division, Directorate General of Human Rights, Strasbourg, France. Retrieved from https://rm.coe.int/1680596143

Council of Europe. (n.d.). Sex and Gender - Gender Matters. Retrieved October 3, 2023, from https://www.coe.int/en/web/gender-matters/sex-and-gender

CRISP (2009). Reaching Rural Women: Designing Programs that Meet Their Needs through a Consultative Design Process. Retrieved from www.reachingruralwomen.org.

Dash, H. K., Jeeva, J. C., Sarkar, A., Mishra, S., & Singh, A. (2015). Strengthening Gender Perspective in Agricultural Research & Extension. Compendium: ICAR Sponsored Short Course on 'Strengthening Gender Perspective in Agricultural Research and Extension,' organized at ICAR-CIWA, Bhubaneswar during 1st-10th September 2015.

Denmark, F. L., Rabinowitz, V. C., & Sechzer, J. A. (2005). Engendering Psychology: Women and Gender Revisited (2nd ed.). Pearson Education New Zealand.

Distefano, F., & de Haan, N. (2018). Gender Analysis in Livestock Management and Interventions in East Africa: Training Report. Report of the FAO Training "Gender and Livestock Development in East Africa," 28-30 May 2018, Nairobi, Kenya.

Dumas, S. E., Maranga, A., Mbullo, P., et al. (2018). "Men Are in Front at Eating Time, but Not When It Comes to Rearing the Chicken": Unpacking the Gendered Benefits and Costs of Livestock Ownership in Kenya. Food and Nutrition Bulletin, 39(1), 3-27. doi:10.1177/0379572117737428.

Etaugh, C. A., & Bridges, J. S. (2017). Women's Lives: A Psychological Exploration (4th ed.). Routledge New York.

European Commission. (1998). One Hundred Words for Equality: A Glossary of Terms on Equality between Women and Men. Directorate-General for Employment & Social Affairs, Publications Office, European Commission.

European Commission. (2001). Communication from the Commission to the Council and the European Parliament — Programme of action for the mainstreaming of gender equality in Community development co-operation, COM(2001) 295 final. Retrieved from http://www.europarl.europa.eu/sides/getDoc.do?pubRef=-//EP//TEXT+REPORT+A5-2002-0066+0+DOC+XML+V0//EN

European Institute for Gender Equality (2019). Gender Mainstreaming Platform Gender Analysis. Publications Office of the European Union,Luxembourg. Retrieved from https://eige.europa.eu/gender-mainstreaming/tools-methods/gender-analysis?language_content_entity=en

FAO. (2011a). The State of Food Agriculture: Women and Agriculture, Closing the Gender Gap for Development. Rome: Food and Agriculture Organization of the United Nations, Rome.

FAO. (2011b). Gender in Agriculture Sourcebook. Food and Agriculture Organization of the United Nations, Rome.

FAO. (2011c). The role of women in agriculture. Prepared by the SOFA Team and Cheryl Doss. ESA Working Paper No. 11-02, Agricultural Development Economics Division. Food and Agriculture Organization of the United Nations, Rome. Retrieved from https://www.fao.org/3/am307e/am307e00.pdf

FAO. (2018). World Livestock: Transforming the livestock sector through the Sustainable Development Goals. Food and Agriculture Organization of the United Nations, Rome. https://doi.org/10.4060/ca1201en

Farnworth, C. R., Kantor, P., Kruijssen, F., Longley, C., & Colverson, K. E. (2015). Gender integration in livestock and fisheries value chains: emerging good practices from analysis to action. International Journal of Agricultural Resources, Governance and Ecology, 11(3/4),262.doi:10.1504/ijarge.2015.074093

Febrina, P. (2022). Taking the Bull by the Horns: Gender Analysis in a Cattle Project in Indonesia International Development, Community and Environment (IDCE), 257.

Feldstein, H. S., & Poats, S. V. (1989). Working together: Gender analysis in agriculture. Vols. 1 and 2. West Hartford: Kumarian Press.

Freestone, K., Remnant, J., & Gummery, E. (2022). Gender discrimination of veterinary students and its impact on career aspiration: A mixed methods approach. Vet Rec Open, 9(1), e47. doi:10.1002/vro2.47.

GDPRD. (2010). Gender and Agriculture. Platform Policy Brief No.3 ,Global Donor Platform for Rural Development. Bonn, Germany

Gender Inequality in India. (2023). Wikipedia. Retrieved from Gender inequality in India - Wikipedia

GLAD.(n.d.).Why Livestock Matter. Global Livestock Advocacy for Sustainable Development, International Livestock Research Institute (ILRI),Naroibi. Retrieved from https://whylivestockmatter.org/gender

IFAD (2010). Gender and Livestock: Tools for Design – Thematic Paper. International Fund for Agricultural Development, Rome. Ret

LBSNAA (2017), Training Manual for Trainers on Gender Responsive Governance, National Gender Centre, Lal Bahadur Shastri National Academy of Administration, Mussoorie-248179, Uttarakhand, India. Retrieved from https://www.lbsnaa.gov.in/lbsnaa_sub/upload/uploadfiles/files/NGC/Publications/Training%20Manual%202019-%20NGC-%20Report%20No%204_6th%20draft.pdf. (Accessed March 9, 2024).

Leach, F. (2003). Practising gender analysis in education. Oxfam Skills and Practice, Oxfam, Oxford. Retrieved from https://policy-practice.oxfam.org.uk/publications/practising-gender-analysis-in-education-115400.

Learning Partnership. (n.d.). Session 4: Equity versus Equality. Retrieved from https://learningpartnership.org/sites/default/files/resources/pdfs/Session% 204%20Equity%20versus%20Equality.pdf.

Lorber, J., & Moore, L. J. (2007). Gendered bodies: Feminist perspectives. Los Angeles: Roxbury.

MANAGE. (n.d.). Capsule Module on Gender Budgeting. National Institute of Agricultural Extension Management, Rajendranagar, Hyderabad - 500 030. Retrieved from https://krishivistar.gov.in/html/Doc/CapsuleModule.pdf.

March, C., Smith, I., & Mukhopadhyay, M. (1999). A guide to gender analysis frameworks. Oxfam, Oxford. Retrieved from https://policy-practice.oxfam.org.uk/publications/a-guide-to-gender-analysis-frameworks-115397.

Mastercard (2022). The MasterCard Index of Women Entrepreneurs. Retrieved from https://www.mastercard.com/news/media/phwevxcc/the-mastercard-index-of-women-entrepreneurs.pdf. (Accessed March 8, 2024).

McDermott, R., & Hatemi, P. K. (2011). Distinguishing sex and gender. PS: Political Science & Politics, 44(1), 89-92.

Miller, B. A. (2019). The gender and social dimensions to livestock keeping in South Asia: Implications for animal health interventions. Gates Open Research, 3(490), 490.

Ministry of Women & Child Development, Government of India. (n.d.). FAQ on Gender Budgeting. Retrieved from https://wcd.nic.in/sites/default/files/GB%20Flyer.pdf.

Muehlenhard, C. L., & Peterson, Z. D. (2011). Distinguishing between sex and gender: History, current conceptualizations, and implications. Sex Roles, 64, 791-803.

Mukherjee, A., Singh, P., Rakshit, S., Priya, S., Burman, R. R., Shubha, K., & Nikam, V. (2019). Effectiveness of poultry-based farmers' producer organization and its impact on livelihood enhancement of rural women. Indian Journal of Animal Sciences, 89(10), 1152-1160.

Njuki, J., & Sanginga, P. (2013). Gender and livestock: Issues, challenges, and opportunities. Brief/International Livestock Research Institute (ILRI),Naroibi.

Njuki, J., Waithanji, E., Lyimo-Macha, J., Kariuki, J., & Mburu, S. (Eds.). (2013). Women, livestock ownership and markets: Bridging the gender gap in eastern and southern Africa. Routledge. ISBN 9781138377103 London https://doi.org/10.4324/9780203083604

NSPDT. (2017). National Smallholder Poultry Development Trust, Annual Report Smallholder Poultry. Retrieved from https://nspdt.org/smallholder-poultry/.

Panda, A. K., Kumar, A., Sahoo, B., & Pattanaik, S. (2021). Gender dimensions of livestock farming in India. VigyanVarta, 2(9), 49-52.

Panda, A. K., Kumar, A., Sahoo, B., & Tanuja, S. (2016). Empowering farm women through livestock and poultry intervention. Compendium: ICAR Sponsored Short Course on "Empowering Farmwomen Through Livestock and Poultry Intervention," organized at ICAR-CIWA, Bhubaneswar during 21-30 November 2016.

Parker, A. R. (1993). Another point of view: A manual on gender analysis training for grassroots workers. United Nations Development Fund for Women UNIFEM, New York.

Patel, S. J., Patel, M. D., Patel, J. H., Patel, A. S., & Gelani, R. N. (2016). Role of women gender in livestock sector: A review. Journal of Livestock Science, 7, 92-96.

Queensland Government. (2009). Gender analysis toolkit. Office for Women Queensland Government Brisbane.

Quinn, S. (2009). Gender budgeting: Practical implementation handbook. Directorate General of Human Rights and Legal Affairs, Council of Europe.

Ramkumar, S., Rao, S. V. N., & Waldie, K. (2004). Dairy cattle rearing by landless rural women in Pondicherry: A path to empowerment. Indian Journal of Gender Studies, 11(2), 205-222.

Ramsak, A. (n.d.). Training manual for gender mainstreaming and analysis: Building capacity for agriculture research for development (R4D) and innovation. Ekvilib Institut, Republic of Slovenia.

Rangnekar, S. D., Vasiani, P., & Rangnekar, D. V. (1992, November). Women in livestock production in rural India. In Proceedings of 6th AAAP animal science congress (pp. 271-286).

Rathod, P. K., Nikam, T. R., & Landge, S. (2011). Participation of rural women in dairy farming in Karnataka. Indian Res. J. Ext. Edu. 11(2): 31-36.

Retrieved from http://www.ekvilib.org/wp-content/uploads/2017/06/00_Introduction_to_the_Manual.pdf

Rota, A., Sperandini, S., & Hartl, M. (n.d.). Gender and livestock: Tools for design - International Fund for Agricultural Development Rome.

Sharma, J. K. (2014). Understanding the concept of sensitisation in humanities and social sciences: An exploration in philosophy of mind. International Journal of Scientific Research, 24(90), 380-400.

Sharma, P., Kaur, S., & Kumari, V. (2023). Mainstreaming gender concerns in agriculture & allied sector (e-book). Punjab Agricultural University, Ludhiana & National Institute of Agricultural Extension Management, Hyderabad, India.

SIDA (2015). Gender analysis — Principles & elements. Swedish International Development Cooperation Agency Sundbyberg, Sweden. Available at https://www.sida.se/English/publications/159386/gender-analysis—principles—elements/.

Simns-Reldstein, H., & Rasheed, S. V. (1989). Working together: Gender analysis in agriculture, Volume I: Case studies. Kumarian Press.

Singh K. and Schiere, J.B. (1993). Feeding of ruminants on fibrous crop residues. Aspects of treatment, feeding, nutrient evaluation, research and extension. Proc. of a workshop, 4-8 February 1991, NDRI-Karnal, ICAR, New Delhi, India. 486 pp.

Social Assessment Report. (2010). Department of Rural Development and Panchayati Raj, Rajasthan.

Source Farnworth, C. R. (2011). Gender-aware value chain development. Paper prepared for Expert Group Meeting "Enabling rural women's economic empowerment: Institutions, opportunities and participation. Accra, Ghana 20-23 September 2011. Retrieved from [Gender-Aware Value Chain Development (un.org)].

Swamy, K., Blümmel, M., Cadilhon, J. J., Colverson, K. E., Reddy, Y. R., & Ravichandran, T. (2014). A gendered assessment of the Mulukanoor Women's Cooperative Dairy value chain, Telangana, India.

Tangka, F. K., Jabbar, M. A., & Shapiro, B. I. (2000). Gender roles and child nutrition in livestock production systems in developing countries: A critical review. Socioeconomics and Policy Research Working Paper 27. International Livestock Research Institute ILRI ,Nairobi

Team, SOFA, & Doss, C. (n.d.). The role of women in agriculture. Food and Agriculture Organization, Agricultural Development Economics Division, FAO, Rome.

Tegbaru, A., Sarapura, S., Odame, H. H., & Fitzsimons, J. (2012). Training manual for gender mainstreaming and analysis. IITA and University of Guelph, Ontario, Canada.

The Hindu (2023). Bridging the Gap: On India and Gender Gap Report. Retrieved from https://www.thehindu.com/opinion/editorial/bridging-the-gap-the-hindu-editorial-on-india-and-gender-gap-report/article66997691.ece.

Thirunavukkarasu, M., & Christy, J. (2002). Socio-economic dimensions of female participation in livestock rearing: A case study in Tamil Nadu. Indian Journal of Agriculture and Economics, 57(1):99-103

Thornton, P. K., Kruska, R. L., Henninger, N., Kristjanson, P. M., Reid, R. S., Atieno, F., Odero, A., & Ndegwa, T. (2002). Mapping poverty and livestock in the developing world. International Livestock Research Institute, Nairobi.

UN-Habitat. (n.d.). Strategies and tools for Gender budgeting in Gender budgeting handbook for Government of India ministries & departments. United Nations Human Settlements Programme,Nairobi Kenya

UNDP. (2007). Gender mainstreaming a key driver of development in environment & energy. United Nations Development Programme,New York.

UNDP. (2008). Training of trainers manual on gender mainstreaming of disaster risk management. United Nations Development Programme, New York Retrieved from https://www.undp.

org/india/publications/training-trainers-manual-gender-mainstreaming-disaster-risk-management.

UNFPA. (n.d.). Frequently asked questions about gender. United Nations Population Fund,New York.

Unger, R. K., & Crawford, M. (1993). Sex and gender—The troubled relationship between terms and concepts. Psychological Science, 4, 122–124.

UNICEF. (n.d.). Gender glossary: Terms and concepts. United Nations International Children's Emergency Fund East Asia & Pacific Regional Office.

UN Women. (2020). Gender mainstreaming: A global strategy for achieving gender equality and the empowerment of women and girls. UN Women – Headquarters. Retrieved from https://www.unwomen.org/en/digital-library/publications/2020/04/brochure-gender-mainstreaming-strategy-for-achieving-gender-equality-and-empowerment-of-women-girls.

USAID. (2013). Integrating gender equality and female empowerment in USAID's program cycle. ADS Chapter 205, United States Agency for International Development, Washington DC.

USAID. (n.d.). Gender and livestock brief. United States Agency for International Development. Washington DC.

Vijayalakshmy, K., Chakraborty, S., Biswal, J., & Rahman, H. (2023). The role of rural Indian women in livestock production. European Journal of Humanities and Social Sciences, 3(1), 91-98.

Vijayamba, R. (2022). Women's participation in livestock raising—Evidence from NSSO employment and unemployment surveys 1993–94 to 2011–12. In J. K. Sharma (Ed.), Gendered inequalities in paid and unpaid work of women in India (pp. 205-219). Springer, Singapore.

Wood, J. T. (1999). Gendered lives: Communication, gender, and culture (3rd ed.). Wadsworth Publishing, Belmont, California.

World Bank. (2007). Global monitoring report 2007: Millennium development goals: Confronting the challenges of gender equality and fragile states. Washington, DC

World Bank. (2009). Gender in agriculture - Source book. The World Bank, Washington, DC.